Developing Technicians

Successful International Systems

Edited by

James R. Mahoney
and Lynn Barnett

Community College Press®

a division of the American
Association of Community Colleges
Washington, D.C.

The development and publication of this monograph was supported by the National Science Foundation through a grant to the American Association of Community Colleges, grant number DUE-97138868.

The editors wish to acknowledge Jennifer Henderson for her research contributions.

Requests for permission should be sent to
Community College Press
American Association of Community Colleges
One Dupont Circle, NW
Suite 410
Washington, DC 20036-1176
Fax: (202) 223-9390

Printed in the United States of America.

ISBN 0-87117-319-0

Contents

List of Figures v

List of Tables v

Foreword vii

Introduction 1
James R. Mahoney

1

Technical Education in Denmark: The "Sandwich" Model 13
Stuart A. Rosenfeld and Cynthia D. Liston

2

Technician Manpower Development in Scotland 35
Nigel Paine

3

Effective Manpower Development Systems in Australia 73
Peter Noonan

4

The Vocational and Technical Education System in Japan: Technicians for Today and Tomorrow 103
Elizabeth J. Teles and Hidetoshi Miyakawa

5

The Technician Manpower Development System in Israel ... 125
Jack L. Waintraub and Haya Adner

6

The German Vocational, Apprenticeship, and Continuing Education System ... 143
Ashok Agrawal

Appendix

Questions from National Science Foundation Invitational Meeting, 1996: Science, Mathematics, Engineering, and Technology Education Programs in Other Countries ... 161

Selected Bibliography ... 164

Index ... 169

About the Editors ... 175

Figures

Figure 3.1 Critical Relationships in Australia's Technical Education System 77
Figure 3.2 Australia's National Training Framework. 79
Figure 3.3 Three Phases of the Development of Australian Skill Standards and Qualifications. 81
Figure 5.1 Israel Electronics Industries Sales. 126
Figure 5.2 Israel Electronics Industries Employment and Manpower Structure 127
Figure 5.3 Comparison of Israeli and International Investment in Research and Development. 128

Tables

Table 1.1 Relationship among Partners in Developing Programs in Denmark 24
Table 1.2 Curriculum for Electronics Technician by Type of Course and School Period in Denmark 29
Table 1.3 Typical Courses for Electronics Technician in Denmark 30
Table 2.1 Relationship among Education Partners in Scotland 40
Table 2.2 Scotland's National Certificate Programme Module Groupings 43
Table 5.1 Typical Electronics Technology Program in Israel: Hours per Semester. 138

Foreword

In the United States we tend to look inward, to see ourselves as apart from the rest of the world, often assuming that we can find creative solutions to any problem without outside help. However, the now-familiar pressures toward globalization have diminished the value of that attitude. In important ways, nations around the globe are similar to one another, with common concerns about such issues as the development and maintenance of economies, the health and safety of populations, and the education and training of young and old.

With the help of advanced communications, answers to common problems are being developed. Solutions no longer appear to be unique to one nation, but, rather, to combine various approaches that fit the particular circumstances of a single nation. The best ideas from around the world are gathered, adapted, and applied in individual settings. The process varies from country to country, but the essential components appear to be eclectic. In short, nations are learning from one another and working together.

The purpose of this monograph is to contribute to this problem-solving process. The monograph presents snapshots of national systems for training technicians from six developed countries with reputations for excellence in educating and training their workforces. Essential components of each of the systems are highlighted, and some of the problems and trends associated with them are noted. We hope that readers will not only find the material interesting but also investigate these systems further with the aid of the bibliographies provided. Most important, we hope readers will consider adapting what is best in these models to create improvements in the American system—locally, regionally, and nationally.

This work is one of the activities supported by a grant awarded by the National Science Foundation to the American Association of Community Colleges. Not enough can be said in appreciation of NSF for its special support of community colleges over the past decade. It has stimulated AACC and our colleges to make extraordinary gains in science, mathematics, engineering, and technology education for the benefit of our students and our national economy.

David R. Pierce
President
American Association of Community Colleges

Introduction

James R. Mahoney

As the United States labor market and the number of jobs that require technical skills grow as a result of the technological revolution, educators and employers are increasingly interested in how individuals will acquire those skills before they enter the workplace. These new technicians, who often are entering emerging technical areas, must have learned the theories and applications of science, mathematics, engineering, and technology fields to be productive workers. Much may be learned from the successful technician development programs operating in other parts of the world.

This was the underlying assumption that motivated more than 30 U.S. experts on technician education who met in Washington, D.C., in 1996 under the auspices of the National Science Foundation (NSF). The experts represented higher education, government agencies, international organizations, professional associations, business and industry, and the public sector. During the two-and-a-half-day discussion, they laid elaborate plans for collecting information about international systems and for sharing the findings. These plans, however, were beyond NSF's scope of interest at that time; a more modest approach to achieving a dimension of the original goal of the meeting was in line with the foundation's concerns.

The more modest approach was to prepare a collection of brief descriptions of technician education systems in countries with reputations for successfully meeting their economic demands for skilled technicians. The descriptions were to focus on several parts of the systems: compulsory presecondary education; compulsory secondary education; the critical juncture at which students make decisions about the academic or career tracks they will take; and the career options available to students.

The importance of U.S. technician education is underscored by certain principles. For their companies to maintain competitiveness in the global marketplace, U.S. technicians and technologists must display at least the same level of skill and competency as their counterparts in other countries. Education and training outcome and performance standards for various occupations in the United States must be similar to or higher than those in other countries, and certifications of experience and competency in specific technical occupations must be comparable. Further, employment patterns of manufacturing and information-related enterprises in the United States and the world require that the competencies of the technical workforce be portable, both within the United States and internationally.

As evidenced by the growing strength of regional trade communities (European Union, North American Free Trade Agreement, Asian-Pacific Economic Community, and General Agreement on Trade and Tariffs), the days of regional and national markets are over. Trade is international; demand for consistent product quality is international; and consistent, high-level technical competency in the workplace is also an international issue. It is to everyone's—every country's, every business's, every worker's—benefit that the best in national technician development systems be identified and applied in countries that wish to compete in this global environment.

This monograph is aimed at practitioners at all levels who have a hand in shaping instructional systems for the development of skilled technicians. The six essays it contains—featuring technician education and training systems in Denmark, Scotland, Australia, Japan, Israel, and Germany—are intended to be thumbnail sketches of the essential components of each country's system, rather than comprehensive treatises on the systems. The authors offer some insights into concerns about maintaining and improving the quality of national approaches, as well as trends pointing in new directions.

TECHNICIAN DEVELOPMENT SYSTEMS

One of the first impressions that jumps out from the details presented in these essays is the similarity of the six systems. Although the applications vary from one country to another, they all appear to be based on compatible visions of the structure, process, and content necessary to reach the goal of preparing world-class technicians. The history and culture of individual countries influence how the visions are translated and where the emphases lie, but the guiding principles are nearly identical. The vision includes some of the following principles:

- Clear national standards are necessary to shape learning.
- Standards are shaped and prepared with the essential input and guidance of relevant industry and employee leaders.
- Standards are reviewed periodically to ensure currency.
- Qualifications and assessment measures are vital to certify student learning and are applied at various steps in learning development.
- Funding is primarily the responsibility of the national government, but in work experience arenas, business and industry contribute to costs.
- Options are built into the system to allow students to move from one track to another.
- Clear learning and training paths are established, beginning early in a student's experience in the system.
- Assessment is performance based, not norm based.
- Education and training institutions are assessed periodically to measure achievement and ensure quality.

Some of the distinctive characteristics of these systems are presented and discussed below. They are grouped in the following seven categories: national approaches, similarity of concerns, administrative structure, curriculum, assessment, continuing education, and teacher preparation.

National Approaches

The first and most obvious characteristic of these six models is that they are all national systems. Unlike in the United States, where state and local entities control the shape and content of education, each of the systems described in this monograph is controlled at the national level. Funding, curriculum or training modules, qualifications and assessment tools and approaches, certification of faculty and training organizations, and practical training options, among other details, come under the aegis of the various ministries of education. In some countries, the Ministry of Labor exerts the same kind of authority for various types of worker retraining, adult technical education, and training for the unemployed. Some flexibility to encourage individual and local creativity and innovation has been built into the systems recently (particularly in Denmark, Australia, and Scotland). But, essentially, these are national systems.

There are both strengths and weaknesses in this design. The strengths include

- national clarity of purpose and direction;
- national recognition of the value of a technical education in building and maintaining a healthy economy;
- consistent performance standards and assessment approaches for various technical fields at various levels in those fields;
- mandated involvement of business, unions, students, and school officials at the national and local levels to ensure training compatibility with the work world;
- mechanisms for periodic review of program components to adjust to changing marketplace directions, processes, and equipment;
- the creation of a mobile workforce whose qualifications are uniformly recognized anywhere in the native country (and, in some cases, in other countries);

- the opportunity to collect uniform national data on student and school performance to help inform necessary improvements in the system.

The weaknesses of a national system include

- a tendency to discourage creativity and innovation at the local level;
- the creation of restrictive career and education paths that permit changes only with difficulty or not at all;
- pressure to force students to make career choices very early in their schooling;
- a strong focus on entry-level preparation for young people, with much less concern for re-entry and upgrade training for adults;
- a tendency to limit advice and counsel on system improvement to a small number of sector representatives.

Each of the national systems discussed in this monograph recognizes the weaknesses noted here (and others), and each is addressing them in various ways. Each of the countries is making strong efforts to support the training and retraining of adults.

Similarity of Concerns

A second obvious similarity that emerges from reading these essays is that each country is addressing the same range of problems and concerns, including

- keeping pace with other economies;
- producing graduates with practical experience, whom employers find valuable and capable of contributing in the workplace;
- training adequate numbers of graduates to meet industry demands;
- controlling the costs of technical training (machinery, laboratories, etc.);

- maintaining an adequate supply of internships, apprenticeships, and other workplace experience opportunities for students;
- meeting the needs of both students and employers;
- collecting reliable and useful data to substantiate the quality of programs;
- convincing students and parents of the value of technician education (rather than an academic path) as a career path;
- ensuring the uniformity of training and education throughout the system.

These are also concerns in the United States.

Administrative Structure

The six systems are organized similarly. Generally, nationally appointed councils or committees are attached to the education ministries to advise on system components. For technician education, the councils or boards are composed of representatives from the various education levels, from government, and from industry and labor. They are officially considered partners in the enterprise. Frequently, one faculty person and a student also serve on a council. In some countries, labor ministries control training for particular segments of the system. Coordination among the ministries is expected. While the councils and committees wield considerable influence, it is ministry administrators who make the final decisions about the work of the schools, colleges, and other training centers. The ministry provides a template by which each training organization operates within the system. Depending on the country, local schools and colleges have varying rights to operate within the framework of the template. For example, Scottish institutions have considerable freedom to maneuver within the framework, while Japanese institutions do not. Essentially, funding is determined by student enrollments and completions.

Curriculum

In each of the countries, curriculum is developed at the national level for use by all training organizations in the system. The curriculum is based on national standards—designed to match skills required at various levels of expertise—developed by the national administration. The curriculum packages take different shapes in each nation, but all of them include details about the duration of the program, learning objectives, skill standards to be reached, assessment approaches, and, in some cases, suggestions on instruction. In some countries (Scotland and Australia, for example), local schools and colleges have the freedom to use resources and pedagogical techniques of their own choosing to guide students to successful completion of qualifications. In Japan, a full-fledged curriculum is provided to local schools and colleges. In each country, the curriculum requires more intense commitment from students than is the case in the United States. In Israel, students studying in the engineering technology program are required to study 40 hours per week; U.S. students engaged in a similar program are expected to study only 24 hours weekly. Also, students are expected to complete programs, and provisions are made to ensure the highest completion rates possible.

Further, in each country, practical training is an integral part of the education program for technicians. Training contracts between students and industries are central to the Danish and German systems, although the designs are different. In Scotland and Australia, flexibility is built into the system through the use of education and training modules or packages. Each technical area is captured in a sequence of modules that build from one level of competency to another. Students may take sequences of modules to qualify at a particular skill level at their own speed and as their time permits.

Assessment

Since each of these training systems is performance based, assessment is a pillar of program structure. Although written examinations are usually given, students must also prove they have the appropriate qualifying skills in order to complete a course or program. Frequently, these performance assessments take the form of solving a common industry problem or designing or developing a new product related to the technical area. This assessment approach is part of every program at every level of competency.

Continuing Education

Continuing education is approached differently by the six countries discussed in this publication. In Israel, lifelong learning is part of the national culture. The country expects that all of its citizens will continue to engage in learning, both personal and career related, all of their lives. The national government is obligated to make such opportunities available. In Australia, continuing education is as important a part of the national education and training system as is the focus on young learners. Scotland's training modules (some 4,000 of them) are available to adults as well as to youngsters in its system of 43 further education colleges. Recently, Scotland instituted a program, called Investment in People, that recognizes organizations of all kinds that provide employees with opportunities for personal and career development. Denmark uses industry-specific training courses through the Ministry of Labour to upgrade the skills of technicians; modules developed by the Ministry of Education are used for the same purpose. German companies invest billions of deutsche marks each year to expand worker skills. Japan appears to have just discovered the importance of providing formal training to incumbent workers. While Japanese companies are now encouraging investment in these activities, most of what exists now focuses on personal or social development.

Teacher Preparation

The record in the area of teacher education is also uneven. In Japan, students know from their freshman year in college that they will be teachers. Most attend one of the schools of education for their training, and there is little reward for improving teaching skills or expanding knowledge in a specialty area. Terminal degrees are not important in the system and few in-service opportunities are provided. In Germany, however, teachers are required to have higher degrees, to have experience outside the classroom, and to participate in targeted training on pedagogical approaches. The system provides liberal opportunities for real-world experiences in teachers' specialties and for further study in their fields. Denmark, Scotland, and Australia require faculty members to have industry experience. Denmark and Scotland offer specialized training for vocational/technical teachers, and each provides teachers with experience in the use of instructional technology. Australia conducts a national teacher recognition program to encourage instructional improvement. Israel's Pedagogy Center for Research and Development trains teachers in the system.

TRENDS AND PROBLEMS

The writers of the essays in this monograph identify many trends and problems that could have been listed directly from commentary on the U.S. system of technician development. These include

1. Reduced involvement of students. A significant concern spotlighted by all the writers is the decreasing number or percentage of students opting for the academic track as opposed to the technical one. This phenomenon is leaving the track with fewer, less-capable students at a time when more highly capable students are needed. Explanations for this decrease include lower birthrates, lower status of technician-level careers, parental influence and guidance, and inadequate recruiting and public relations efforts.

2. Reduced involvement of industry. Situations in several countries suggest that their industries are less willing than before to participate in apprenticeship aspects of the national training program. Explanations for this reduced involvement include structural changes in the industries, downsizing, outsourcing, and uncertainty that the benefits justify their expense. Further, mismatches between the practical experiences that students want and those offered by industry create tensions that discourage industry. Although in most of these countries industry spends billions annually to support train ing of various sorts, some critics believe that industry is not contributing enough.

3. Declining power of unions. Particularly in European countries, economic recessions in the 1990s have reduced the power of unions in the education and training enterprise. The systems are built upon cooperation among the equal authorities of industry and employee groups. With a weaker union, the equation becomes unbalanced and industry influence predominates.

4. Worker mobility as an advantage and a liability. National standards are designed in part to ensure that students emerge from programs with the same level of skill and competency regardless of the institution that provides the training. One result is that students are mobile and can relocate to any part of the country they choose (and sometimes to other countries). A negative result of this mobility is that the local industries that trained students are left shorthanded when students relocate, while other industries and localities reap the benefits of the local industries' training successes. The imbalance can affect the national economy.

5. Producers, not entrepreneurs. Tight, nationally orchestrated training systems tend to develop producers, not entrepreneurs. That is, the systems tend to prepare employee technicians who are highly skilled at generating products in the company line. The systems are much less likely to stimulate the kind of imaginative, creative, and risk-taking person who might strike out to organize a small business. New small businesses are the lifeblood of most economies.

6. Proportion of general and technical education. Many of these countries are struggling to find the right balance between general and technical education. A technical emphasis prepares students for specific careers in particular industries—an approach that may limit graduates' employment options graduates and reduce their capacity to adapt quickly to changes in the workplace. In the short term, the approach favors employers. An emphasis on general education may reduce the contribution made by new employees in the short term, but may enable them to make stronger contributions in the long term. Generally, employers favor a technical emphasis and educators prefer general education. Students appear to choose the general education emphasis because it gives them more options for career choices.

7. Basic skills emphasis. Many of these countries are moving toward an emphasis on basic skills development as a foundation for technician education. These basic skills include problem solving, communication, team building, and creative thinking. This movement is a response to the new demands of the quickly changing workplace and is consistent with what is happening in countries around the globe.

8. Adjusting to rapid change. All of the technician education systems described in this monograph are confronted with the problem of rapid change in the workplace. Processes, equipment, and products are changing so quickly that national ministries find it difficult to adjust curricula, standards, and assessment approaches to match the changes. In a tightly controlled and inflexible system such as Japan's, change is slow, arduous, and costly. Change is easier in systems such as Scotland's and Australia's, which have modularized training programs and allow some local autonomy in adjusting curricula. But rapid change is a problem for all of these countries. A similar problem exists in developing curricula to train students for new occupations or specialties.

9. Limited pathways for students. To varying degrees, each of these countries is becoming more sensitive to student interests. Some of these highly organized systems force students to make career decisions as early as age 14 or 15, with little opportunity afterward for shifting schooling or career direction. With the possible exception of Japan, all of the countries have altered or are altering their systems to permit students to move among the various tracks (including from vocational to academic and from academic to vocational) as their interests and opportunities change. Credits for successfully completed work are carried with students, and, depending on the nature of the credits, are applied to the new program. Still, there is a certain degree of inflexibility in all of these systems.

10. Other problems and trends. A number of other problems and trends are evident in the country reports. They include

- uneven apprenticeship experiences caused in part by differences in the quality of equipment available on-site in training firms;
- conflict between employers and unions concerning training levels, with employers pressing for just enough specific training on-site and unions stressing broad, general training;
- the need for better, more comprehensive management information systems to analyze the quality and achievement of each system;
- improved monitoring of employment-based training;
- better approaches to assessing and crediting informal learning on and off the job;
- more responsive approaches to training immigrants and persons with disabilities;
- stronger public information systems to enhance awareness of and regard for technician education systems.

The similarities and differences between these six systems and the U.S. approach are clear. One of the certain conclusions arising from these essays is the commonality of philosophies and missions expressed in them, as well as the challenges each country faces. The hope is that the details in these essays may help trigger innovations and adaptations in the U.S. system that will lead to improvements.

Chapter 1

Technical Education in Denmark: The "Sandwich" Model

Stuart A. Rosenfeld and Cynthia D. Liston
Regional Technology Strategies, Inc.,
Chapel Hill, North Carolina

INTRODUCTION

Denmark is the southernmost Scandinavian country in Europe. The nation consists of the Jutland peninsula and an archipelago of 406 islands, of which 89 are inhabited. The largest island, Zealand, is home to most of the nation's 5.2 million people, including about 2 million in the greater Copenhagen area. A largely agrarian country until the late 19th century, Denmark is now one of the most industrialized nations in Europe. This transformation started with the introduction of modern agricultural and food-processing techniques that made Denmark a major exporter to the rest of Europe.

Today, close to 22 percent of the workforce is employed in manufacturing, with machinery, transportation equipment, and food and consumer products (including furniture) playing the largest roles. Denmark has a positive balance of trade with both Germany, its largest trading partner, and Japan. Much of its success can be attributed to its ability to develop, adopt, and effectively use advanced technologies, aided by a dense science and technology infrastructure, which includes the 1,000-employee Danish Technological Institute (founded in 1906 as a training institute for industry), a set of applied research institutes, research centers and parks, research universities, and a highly regarded educational system.

The nation spends 6.7 percent of its gross domestic product (GDP) on education. Its technical education system and the workforce it produces are highly regarded among educators in the United

States, visited often by U.S. delegations, and frequently cited as among Europe's leading models for U.S. school-to-work programs, youth apprenticeship initiatives, and skill standards. A 1995 study published by the Organization for Economic Cooperation and Development (OECD) says that "the quality of education is indeed quite high at all educational levels. . . . this well educated workforce is able to operate relatively independently at all levels."

Denmark is highly unionized—more than 80 percent of its workforce (including white-collar professionals) belong to a union. It also has the highest proportion of women in the labor force of any European nation. Organized labor and employer associations both play leading roles in setting policies and standards for educational programs, as well as in delivering education.

MISSION OF DENMARK'S TECHNICAL TRAINING PROGRAM

Much of Denmark's strength in education and training is the result of learning from centuries of tradition and experience. Denmark has a rich history of vocational education and training. This history began with the medieval guilds; continued with technical schools operated by the guilds in the mid-17th century; progressed to 19th- and early 20th-century apprenticeship programs at evening schools; and evolved to its current "sandwiched" classroom and workplace-based diploma, certificate, and degree programs. The highlights of the progressive development of Denmark's system are as follows:

- In the mid-17th century, Denmark's first technical schools were established for clothing production workers, dockworkers, and carpenters.
- In 1870 Denmark had 50 technical schools; by 1910 the number had grown to 170.
- The 1889 Apprenticeship Act allowed the government to regulate the technical training system through syllabuses, texts, and standard examinations.

- The 1921 Apprenticeship Act created a "social partnership" among employers, unions, and schools.
- In 1956, when the Danish economy was booming and skilled labor was in great demand, all restrictions on entry to apprenticeship programs were abolished. The system was also reformed, with classes moved from evenings to daytime and new course requirements added. Colleges also were expected to choose areas of technical specialization.
- In 1977, in response to criticism that students were forced to choose careers too early, to rising unemployment, and to shortages of apprenticeship positions, the Basic Vocational Education Act reorganized technical education to begin with a year (sometimes less) of school-based education for occupational clusters and to offer an alternative path to the traditional apprenticeship.
- Reforms in 1989 streamlined the system, consolidating the number of programs and establishing the foundation for the current system.

According to the Act on Vocational Education and Training of 1989, the goal of Denmark's vocational education system is to

1. Motivate young people to study, and to ensure that all those who want to obtain a vocational qualification have genuine opportunities to choose from a variety of different streams;
2. Provide young people with a basis for future employment, and at the same time contribute to personal development and understanding of society and its development;
3. Meet the needs of the labor market for vocational and general qualifications that are assessed with a view to the development of business production and society, including ongoing changes in the structure of the business sector, in labor market conditions, in workplace organization, and in technology;
4. Provide youth with a solid basis for further education and training.

DANISH EDUCATIONAL SYSTEM: AN OVERVIEW

Almost all education in Denmark is publicly funded, free of charge, and open to everyone. Schooling is compulsory for nine years (with an optional 10th year), and elective thereafter, but nearly everyone continues on to some form of further education. The school year typically lasts 200 days and teaching time, depending on level, varies between 15 and 34 lessons. The vast majority of students with special needs are mainstreamed, and only about 2 percent of students are in special classes. The average pupil-to-teacher ratio is 19 to 1 and the maximum number of students allowed is 28. Given the small size of the country, foreign languages are important, and schools emphasize language skills, with requirements in English and German.

The Danish educational system begins with the primary and lower-secondary *folkeskoles*, after which 95 percent of youth continue on to upper-secondary school at either an academic *gymnasium* or a vocational college (of which there are commercial, technical, agricultural, and health-care colleges). After secondary school, about 15 percent of Danish youth go on to pursue higher (tertiary) education.

Folkeskoles

Compulsory education in Denmark legally begins at age seven, although most children start at age six. About 89 percent of Denmark's young children attend public folkeskoles, which cover primary and lower-secondary grades—the nine years of compulsory education typically from ages 7 though 16. The remaining 11 percent attend private schools, where 80 percent of costs are paid by the government. There is no overall examination at the end of lower-secondary education; however, students may take leaving exams in single subject areas, five of which can be taken at an advanced level. Students themselves decide which, if any, subject exams they will take, and almost all sit for at least some of the 11 authorized exams. In some subjects, students are allowed to repeat exams to improve their scores. In 1992, for example, 96 percent of all pupils passed either the leaving or advanced-leaving exam in mathematics.

During grades seven through nine (and 10 if a student stays in school), students receive career guidance and can choose to spend up to a week at a time in various businesses. Schools offer group and individual counseling to help students choose the educational path they will follow after folkeskole. Whereas a large percentage of working adults did not pursue education beyond the folkeskole, today about 94 percent of all youth continue on to upper-secondary education.

Secondary Education

The upper-secondary education system starts at the conclusion of compulsory education. There are two tracks: About 45 percent of all youth attend a language or mathematics gymnasium for a general academic education that prepares them to attend university; about 50 percent pursue a vocational course of study in one of the 58 technical colleges, 54 commercial colleges, or 30 agricultural, health, or social care colleges. Choosing the vocational track does not preclude attending university later.

Technical Training

Entry into a vocational education and training program is open to anyone who has completed compulsory education; no entrance examination or certificate (diploma) is required. However, entrance into some programs may be restricted as a result of changes in the labor market that portend decreased job opportunities for certain careers.

Within vocational education and training (VET), students can choose one of two paths. About two-thirds select one of 87 "basic" vocational programs—all of which have sub-specialties—that typically take three or four years to complete and after which most graduates enter the workforce. (The number of occupational programs was reduced from about 300 under the education reforms of 1991.) These students spend approximately two-thirds of their time in the classroom and one-third working in companies gaining practical experience through apprenticeships—popularly called the "sandwich" model. From the beginning, hands-on learning is emphasized over theory. Although the vocational specialty is the primary focus of these programs, general education is also part of all occupational training through its integration into technical courses.

The second vocational track, created on an experimental basis in 1982 and now pursued by about 15 percent of Danish youth after compulsory education (30 percent of all vocational students) is purely school-based and leads to an HTX (higher technical) or HHX (higher commercial) examination. It was established to strengthen the status of technical colleges. HTX and HHX programs are taught in the same technical and commercial colleges where basic vocational programs are taught. HTX and HHX students typically spend their first year (or six months, depending on the program) with basic vocational studies in their specialty. The next two years are spent taking more theoretical curricula than are taken by students in a basic vocational program. After graduation, many of these students go to work directly into the industry they studied while others continue to study at the university level. Both the basic and the HTX or HHX vocational tracks result in a certificate, but students who complete an HTX or HHX program are fully qualified to enter universities or other higher education institutions.

In the 1980s the Danish government significantly expanded the HTX and HHX tracks to their current levels in response to criticism that Denmark's educational system was too rigidly tracking youth at too early an age. Although students still choose between academic gymnasiums and vocational upper-secondary schools quite early (at age 15 or 16), either track can provide entry to higher education. In addition, qualified students in a basic vocational program may transfer to an HTX program after their first year. These reforms have enhanced the image and status of vocational education in Denmark.

Credit Transfer among Institutions

Through another reform, based on an executive order of the minister of education, credits may now be transferred between different types of upper-secondary programs. Once a student passes an examination with a grade of at least 6 out of a possible 13, the credits can be transferred to another program at the same level. Headmasters also are authorized to give exemptions for courses passed outside of upper-secondary education.

OPPORTUNITIES FOR CONTINUING EDUCATION AND TRAINING

Denmark has a separate system—the AMU system—for training or retraining adults. Training semiskilled workers and the further education of workers who completed both basic and HTX or HHX vocational programs are the main functions of AMU programs. The courses offered through the system range from short "worker introduction" classes for individuals who are reentering the workforce to highly technical computer-based production classes for skilled employees. AMU courses are taught in modules to encourage flexibility. Adult learners may choose to take only one module or to take a whole series of modules that results in a certificate recognizing the completion of connected courses. The system is designed so that students may enter and exit with relative ease and continuity.

Dedicated AMU training facilities house some AMU programs, while more advanced skill upgrading is also taught at the same technical and commercial colleges that educate Danish youth. Courses, which are free to students, are paid for by mandatory employee and employer contributions to a vocational training fund, and students receive living stipends while enrolled in full-time programs. In 1994, about 180,000 adult learners (7 percent of the workforce) took part in the AMU system.

With the goal of providing more comprehensive technical education to adults, a parliamentary resolution in 1992 established vocational education and training for those 25 or older, aimed primarily at unskilled working adults. Offered by the technical colleges and AMUs, these programs are based on the same sandwich model as the youth courses. They require a contract with an employer, which ensures participants a salary during the education periods. About 3,000 adults were enrolled in these programs in 1993.

To complement the AMU system and better enable lifelong learning, in 1990 the Danish Ministry of Education instituted the open education system for adults wishing to pursue part-time studies on their own (in contrast to AMU programs, which are usually arranged through employers). The open education program gives adults access to all courses available to youth, including the same credit status. In 1993, about 140,000 adults (5.5 percent of the workforce) enrolled in open education courses. Grants from the Ministry of Education cover about 80 percent of a student's costs.

The Ministry of Labour also offers enterprise-oriented courses, which are tailored to meet the job-specific needs of individual businesses that are not met through ordinary adult training. These programs result in completion certificates for graduates.

TECHNOLOGY-BASED LEARNING

A nation of islands, Denmark spends considerable resources to improve the quality of and access to education through technology. The effort is spearheaded by the Danish Center for Technology-Supported Learning, which funds pedagogical research and innovative pilot projects that use educational information and communication technologies. The pilot projects emphasize adapting technologies for use in teaching; creating distance learning networks for asynchronous delivery of courses; developing resource centers on computer-based technologies; producing in-service training for teachers; creating multimedia curricular content; and promoting partnerships between the education and noneducation sectors.

NATIONAL ADMINISTRATIVE STRUCTURE

Denmark's vocational and education training system operates under two administrative authorities. The formal and credentialed educational programs mainly targeted to youth, available in the technical and commercial colleges, operate under the Ministry of Education. Retraining and further education courses targeted at the existing workforce (the AMU system) operate under the auspices of the Ministry of Labour.

Each technical and commercial college has a school governing board composed of 6 to 12 members representing local employers, unions, city and county governments, students, and college staff. The board governs the college, which is considered independent, approving budgets and deciding what types of education the school should offer. Regional economic development is an explicit role for vocational colleges in Denmark, and the fields of study the colleges offer are supposed to reflect local economic conditions.

Other local involvement in technical and commercial colleges comes in the form of local education and training committees, which advise the various curricular programs and promote cooperation between the college and the labor market. These committees help translate national regulations into concrete classroom activities and obtain training positions for students in industry.

Although each college controls its daily management, the national government exercises significant control over the content of courses and requirements for vocational programs. The national Department of Vocational Education and Training, which resides within the Ministry of Education, is advised by a 20-person Vocational Training Council that is responsible for regulating and rationalizing the system. The minister of education appoints the chair of the council; the Danish Employers' Confederation and Danish Confederation of Trade Unions appoint eight members each; and the ministers of education and research, labor, and industry appoint one member each. The council makes recommendations about regulations for the structure of training, content, and examination procedures; approves new initiatives or experiments; and regulates the rights of students. Although the minister alone makes final ruling, he or she is required to consider the council's recommendations. The council can set up initiative and coordination committees for specific business sectors to draw attention to the need for new programs, but the industry organizations represented must bear the costs.

National-level trade committees formed jointly by labor and business organizations are responsible for the program curricula (content, duration, objectives, and examination standards) for their industries or occupations. These committees, again financed by their own organizations, must approve colleges as qualified training institutions for their industries.

The Adult Vocational Training System (National Labour Market Authority, or AMU system) falls under the Ministry of Labour, advised by a Training Council for Adult Vocational Training, with a composition similar to that of the Vocational Training Council except that the minister of labor appoints the chairman. The minister forms training committees for unskilled and semiskilled workers, pre-employment training, further training of skilled workers, and contact committees for supervisors. A chair is appointed for each committee. The unions and employer associations appoint the committees' members (eight each to each committee). These committees must approve all training programs and curricula and make recommendations for the allocation of resources.

Although the number of technical institutions in the Danish system is large (more than 100, with 58 dedicated to technical curricula), specialization among individual schools minimizes duplication of effort. Since the reforms of 1956, colleges have been expected to select areas of concentration based on the needs of their local economies. For example, EUC-Syd in Sønderborg specializes in electronics, Slagteriskolen in Roskilde serves the meat-processing industry, Metalindustriens Fagskole in Ballerup concentrates on metalworking, and the Technical College of Jutland in Hadsten educates for the crafts and entrepreneurship.

SOCIAL PARTNERS: ROLES OF PUBLIC AND PRIVATE SECTORS

Goals for education and training are set by Denmark's social partners—government, business, and labor. Even Danish educational standards are set by these social partners, not by the Ministry of Education acting alone. Thus, employers have a very real role in education and training. Colleges also are guided by trade committees in particular specialties: One trade board and one industry association is assigned to each education program offered. The iron and metal workers union, for example, represents labor for the electronics technician programs.

Procedures and structures for designing and altering technical programs are formal and comprehensive. Plans are drawn up according to the vocational education laws, the public is given notice, and plans of study for each school period (typically 10 weeks of classroom time) are drawn up by trade committees working in cooperation with local education committees. Table 1.1 *(page 24)* shows the responsibilities of each of the actors in a typical curriculum.

Because much of education takes place at the worksite, employers also have a formal role in the education process. Under their training agreements, students are expected to try different positions within companies in order to be exposed to a variety of challenges. Technical education certificate programs in Denmark are very narrowly defined, and students learn about industry and the economy informally. There is no structured interaction with other students during workplace assignments.

ASSESSMENT

Upon completing a vocational program, students face a final assessment. Depending on the program, this may be a journeyman's performance-based competency examination, a school-based examination, or, more commonly, a combination of the two. Because industry standards and certifications are set by national trade committees and other

Table 1.1 Relationship among Partners in Developing Programs in Denmark

Entity	Legal Foundation	Roles	Relation to College
Vocational education and training law	Introduces and publishes notices	Sets general parameters	Establishes schools
Ministry of Education	Confirms	Sets general scope	Supervises system
Vocational Education Council	Recommends implementation	Proposes subjects	n/a
National Trade Committees	Review	Select content, study areas, and scope	n/a
Local Education Committees	Monitor	Propose and manage optional courses	Supervise and monitor programs
School Boards	Market programs	Involved only in electives	n/a

regulators, these trade committees determine the performance level required to pass in each technical field and issue a certificate that allows those who pass to join a union and qualifies them for regular worker pay scales.

Generally, one employer representative and one employee representative select an exercise for a student to complete to qualify for graduation, often some form of real-life problem such as an equipment malfunction. Since the employer is responsible for the student passing the practical exams, students remain at their jobs at full pay until they pass.

Until 1991, schools could elect not to give tests that measure general educational competencies so as not to penalize practical-oriented students, but regulations now require written exams of all students in technical programs. The Ministry of Education sets the general education performance levels, and the schools administer and grade the tests.

TEACHER PREPARATION

Teachers of an occupational trade in a technical college must have at least five years of experience in their occupation. Teachers of more general subjects may be hired after only two years of experience. In addition to a basic vocational certificate in his or her occupational area, a teacher must also have relevant further education (AMU) or higher technical qualifications. During their first two years of teaching, all teachers complete six months of pedagogical training, specifically designed for vocational teachers, provided by the State Institute for the Educational Training of Vocational Teachers. This training includes both classroom experiences and teaching "tutors" who observe and guide new teachers in the classroom.

While industry experience is still important, in the past 20 years the trend has been to hire teachers with greater levels of education than had been the case previously. According to a 1991 report, teachers are usually—but not always—better paid than their counterparts in their fields.

THE "SANDWICH" MODEL FOR BASIC TECHNICAL PROGRAMS

The majority of Denmark's technicians are trained through the basic vocational path. The word *basic* can be misleading because these paths are composed of intense and high-quality technical programs structured as Denmark's unique version of an apprenticeship program. It is popularly called the "sandwich" model because students alternate extended periods of classroom-based education with extended periods of workplace-based education. This approach differs from the German or Austrian model, in which each week includes both work experience (three-fourths of the total education time) and classroom instruction (one-fourth of the total education time). Every company in Denmark with 10 or more employees is required to accept apprentices.

Technical programs can begin at school, on the job, or with an introductory generic technical course. Typically, the first sandwich "layer" is in the college and lasts between six months and a year. After that, the student alternates periods of 20 weeks on the job with 10 weeks in school for three years. Schoolwork becomes increasingly specialized over time. Programs take between three and four years to complete; graduates are typically 18 or 19 years old.

For the workplace portion of the training, students are expected to secure a training agreement with a local firm. The agreement covers all in-company training, including a journeyman's test, if one exists for the occupation. In some instances, two or more businesses jointly agree to host a student, which makes it easier for smaller firms to participate in the national training system. Trade committees must approve a business as a training site before it can host students.

The costs of education and training in workplaces are borne by taxes, but employers are required to pay students' wages (at least half of minimum wage, which is about $12 per hour) during both workplace and classroom training, as set by collective bargaining agreements. Every employer above a certain size must contribute to a training fund. Firms that accept apprentices are reimbursed from this fund for 90 percent of their apprentices' wages during their in-school periods. Employers must submit formal reports that become part of a student's portfolio, which follows each student throughout his or her schooling, including formal communications between the employer and the school.

Some students fail to find their own placements. In these cases, the school tries to provide a simulated workplace experience. Students who cannot find sponsoring employers are not precluded from gaining industry experience. Many of these students are still able to contract for short periods of work—usually six months—with companies. The remaining part of their education is divided between the classroom and the college's simulated place of employment. Achieving the proper balance between the number of students seeking placements and the number of firms offering slots is a continual struggle for the Danish system. It is exacerbated by fluctuations in the economy and by labor market changes in which new jobs—such as computer programming—increase in number while old ones decrease. Often, the problem is one of finding the right match, not the total number of placements available.

Example of a Typical "Basic" Program for Electronics Technicians

Erhvervs Uddannelses Center-Syd (EUC-Syd) Technical College in Sønderberg in southern Denmark officially became Denmark's primary electronics college in 1956. At that time, a vocational education reform act centralized technical education and required each college to develop occupational specialties. The electronics industry grew dramatically in southern Denmark in the 1980s, due in large part to the electronics expertise at both EUC-Syd and the local polytechnic. The college added an automation technician program in 1978 and a data mechanics program in 1984. The college has a 19-person board of directors. Eight represent employers, eight represent employees, two are mayors, and one is a resident of the region. The board meets four times per year and has special committees that also meet at least four times per year. A six-person local committee appointed by the college's board advises the electronics program.

EUC-Syd at Sønderborg offers basic vocational studies (certificate), advanced engineering programs (HTX), short courses for industry (about 40 percent of the college's total activity), and management courses. All faculty members are required to have industry experience. They remain knowledgeable about the latest methods and technologies through their industry courses and frequent interaction with industry managers. Faculty constantly work directly with industry representatives to develop new courses to meet changing work requirements, and generally produce four or five new courses each year. Examples of special programs to meet regional industry needs are mobile phones/telecommunications, medical equipment, and electronic security.

Basic Electronics Path

The basic vocational program (sandwich model) for the electronics technician program begins with six months of general schoolwork as a broad introduction to the many areas of specialization. It is followed by sandwich "layers" that consist of four 10-week and two 5-week school periods that are separated by about 20 weeks of work experience. The first two 10-week classroom periods are generic to clusters of occupations and not separated according to technical specialty. Sample courses during these periods include physics, basic electronics, measurement and documentation, power supplies, and analog and digital devices. Beginning with the third school period, courses are divided between fundamental courses, general vocational area courses, specialty courses, and optional courses, according to the schedule in Table 1.2. *(right)*. Instruction is intense, requiring 37 hours of classroom work each week.

Examples of fundamental courses are physics, math, personal computer technology, and digital impulse technology. Examples of general courses are basic electronics, analog systems, digital systems, and microcontrollers. Examples of specialty courses are high-frequency technologies, high-frequency systems, and telecommunications services. Table 1.3 *(page 30)* shows typical courses for electronics technicians. Since each student is expected to find a work placement, one advantage of EUC-Syd's program is that it is located in a region with a cluster of electronics companies. Most students—more than 90 percent of students in electronics programs, about 80 percent of students in automation programs, and about 55 to 60 percent of students in the more competitive data and computer programs—are able to

Table 1.2 Curriculum for Electronics Technician by Type of Course and School Period* in Denmark

	Duration in Weeks				
Type of Course	**3d**	**4th**	**5th**	**6th**	**7th**
Fundamental	2.2	3.2	3.3	0.8	0.8
General area	5.5	4.5	2.8	0	0
Specialty (Electronics)	0	0	3.9	4.2	4.2
Optional	2.3	2.3	0	0	0

**Note:* A period equals approximately 10 weeks.

acquire training agreements with businesses. Students who fail to secure contracts with employers are assigned to simulated work experiences in the school, which are generally supplemented by some short periods of workplace education.

Although on the surface the course content appears narrowly technical, many economic and social aspects of technical occupations are integrated into the curriculum and work experiences. For example, during the sixth school period, students are taught about connections between society, labor markets, and work, and they are expected to review literature and write compositions demonstrating that they understand economies and labor markets and how changes and trends might affect their work. Quality control and marketing issues are also integrated into the curriculum. However, specific business courses on production management or distribution, for example, are absent even as electives because they are offered in the commercial, not the technical, colleges.

SYSTEM EFFECTIVENESS AND KEY ISSUES

While Denmark's education system produces highly educated technicians, there are mounting concerns about certain trends and prospects

Table 1.3 Typical Courses for Electronics Technician in Denmark

School Period	Courses
3	Physics Basic Electronics Fundamentals of Analog Circuitry Measures and Documentation Testing
4	Servicing of Digital Equipment Optics Programmable Controller Technology Microcontrollers
5	Fundamentals of High-Frequency Technology High-Frequency Theory and Systems Sales and Service
6	Labor Market Conditions High-Frequency Transmission Service and Measurement
7	Labor Market Conditions Servicing Computer Equipment Communications Equipment Journeyman's Exam Entrance Exam

for the system's future. One recent critique of the Danish education system notes a downward trend in technical education that, the authors contend, is caused by three things: parents encouraging their children toward academic studies; fewer businesses taking part in the apprenticeship program (an effect of the recession of the late 1980s, when some businesses eliminated their apprentice positions); and a decreasing birthrate and reduced youth cohort—from 88,000 youth two decades ago to 59,000 predicted for 1998.

One pressing issue in Denmark is an international trend that affects almost every industrialized nation: Industrial occupations are still accorded low status by Danish parents despite the high demand and potential for high income. Many parents are unaware of how technology is altering the workplace and raising required skill levels, and they counsel their children away from vocational and technical programs and toward universities.

Another issue is the increasing numbers of mismatches between students' interests and available apprenticeship positions. Although demand for apprentices in Denmark remains high, too many students are choosing commercial fields where there is an excess of workers. Further, students are increasingly selecting the gymnasium track (more than 60 percent), leaving the lowest-ability students enrolled in the technical vocational tracks. According to a recent study, after 14 years, 30 percent of gymnasium leavers have not completed a higher education program.

The result of these two trends is lower enrollments in technical programs, lower-ability students applying for those programs, and fewer real work experiences for those who do enroll. If these trends continue and the system continues to use full-time equivalent formulas for its funding, the staffing of technical schools will be reduced from 19,000 to 12,000 over the turn of the 21st century. In addition to these trends, Denmark's economy faces its own set of challenges—for example, increasing levels of outsourcing among the country's largest electronics industry employers and the difficulty that local suppliers have to remain competitive.

Another issue is the ability of the educational system to meet the technology requirements of the changing workplace. Because a large portion of training takes place in the workplace, students learn how to use the machinery and processes that industry is using, unlike educational systems where the machinery taught in classrooms and laboratories lags behind that used by firms. Industry in Denmark is by-and-large technologically advanced, giving students experience using advanced equipment. However, not all firms are at equal technological levels, meaning the experiences some students gain in their training companies is not as advanced as those gained by others.

Further, Denmark's workers are highly mobile. An analysis performed by the Danish Technological Institute found that 1.2 million of the country's 2.9 million workers changed jobs and status in one year. This taxes the nation's educational system because its programs are narrowly focused. One response to this issue (and to business restructuring) has been a drastic lessening of education specialization, with the number of programs being reduced from about 300 in the 1980s to fewer than 90 today. Also, the open education system established in 1990 helps remedy worker mobility issues by making retraining more accessible and giving individuals more responsibility for their continuing education.

Turning to the content of Danish educational programs, there is widespread belief among industry leaders that the system needs to spend more time teaching communication skills, work organization, quality management and control, and administrative tasks. This was argued in a 1991 study that examined future training and skill needs, education and training efforts, and mechanisms for cooperation between education and industry. The final report, *Kompetence 2000* (Ministry of Education and Research 1991), which was supported by a large number of Danish organizations, urged the government to address its recommendations. The government is responding.

There is concern about the ability of the educational system to generate and prepare entrepreneurs. Some critics contend that changes in the culture (for example, passive entertainment through television and highly structured playtime) and the rigidity of the educational system are counter to entrepreneurial habits. The Danish system (like most other Western systems) graduates students best prepared to become employees, not employers. Yet downsizing and outsourcing are making business start-ups more important to an economy's health. In response, the Ministry of Education declared 1996 the Year of the Entrepreneur, and produced a report called *A Coherent Educational Strategy on Entrepreneurship* that recommends changing education so that it nurtures entrepreneurial, innovation, and independence "cultures." These changes, now being implemented, involve new pedagogies, revamped teaching materials and guidance, and in-service training of teachers. The pedagogical principles include giving learners more responsibility for the learning process; making lessons more project- and problem-oriented, experience based, and multidisciplinary; and tailoring teaching to learning styles.

In addition, even though Denmark has been a very culturally homogeneous society, an increasing number of immigrants are arriving who require that greater educational investments be made in order for them to be able to adapt to the society and economy. This is likely to present a new kind of challenge to the educational system.

Finally, Denmark must avoid resting on its laurels and relying on its past strengths if it is to continue to compete economically on the basis of the technical competencies of its workforce. It must continually adapt and improve its educational system.

ABOUT THE AUTHORS

Stuart A. Rosenfeld is president and founder of Regional Technology Strategies, Inc., a nonprofit organization in Chapel Hill, North Carolina, that researches, designs, implements, and assesses development strategies in the United States and Europe. His work has emphasized the effect of collaboration, community colleges, and technology diffusion on rural development. Previously, he was deputy director of the Southern Growth Policies Board, an interstate compact comprising 14 southern state governments, where he founded and directed the board's Southern Technology Council.

Cynthia D. Liston is a policy associate with Regional Technology Strategies, Inc., in Chapel Hill, North Carolina, where her policy focus is technology-based economic development and technical education. She also manages the activities of the Trans-Atlantic Technology and Training Alliance, a consortium of leading technical colleges in the United States and Europe.

BIBLIOGRAPHY FOR CHAPTER 1

Brauner, Bent, and Casper Syskind. 1996. *Education and Training at Vocational Colleges in Denmark*. Copenhagen: Danish Ministry of Education.

Council of Chief State School Officers. 1991. *European Lessons from School and the Workplace*. Washington, D.C.: Council of Chief State School Officers.

Industrial Research and Development Advisory Committee of the European Commission. 1994. *Quality and Relevance: The Challenge to European Education*. Brussels: European Commission.

Minister of Education. 1994. *Danish Youth Education: Problems and Achievements. Report to OECD*. Copenhagen: Danish Ministry of Education.

———. 1994. *Technology-Supported Learning (Distance Learning)*, Report No. 1253. Copenhagen: Danish Ministry of Education.

Ministry of Education and Research. 1991. *Kompetence 2000, Hovedrapport*. Copenhagen: Ministry of Education and Research.

———. 1992. *Characteristic Features of Danish Education*. Copenhagen: Ministry of Education and Research, International Relations Division.

———. 1992. *Education in Denmark: A Brief Outline*. Copenhagen: Ministry of Education, International Relations Division.

———. 1996. *A Coherent Educational Strategy on Entrepreneurship*. English Translation. Copenhagen: Ministry of Education and Research.

Minister of Labour. 1992. *The Danish Labour Market Authority: Your Access to Effective Adult Vocational Training-AMU*. Copenhagen: Ministry of Labour.

———. 1997. *Adult Vocational Training*. Copenhagen: Ministry of Labour.

Nielsen, Maj Cecile, and Niels Christian Nielsen. 1997. *Verdens Bedste Uddannelses-System*. Århus, Denmark: Fremd.

Organization for Economic Cooperation and Development. 1995. *Science, Technology and Innovation Policies: Denmark*. Paris: Organization for Economic Cooperation and Development.

Østerlund, Roland. n.d. "School-to-Work Transition in Denmark." Unpublished paper. Copenhagen: Department of Education and Training, Ministry of Education.

Ploughmann, Peter. 1993. "Meeting the Needs for Intermediate Skills and Technical Qualifications: The Danish Case." Paper presented at a European Union Workshop September 15–16, Paris. Taastrup, Denmark: Danish Technological Institute, Human Resources Development.

"Undervisningsplaner for Elektronikmekanikere." 1996. Sønderborg, Denmark: EUC-Syd.

Winther-Schmidt, Erik. 1992. *The Danish Vocational Education and Training Systems in the Single Market*. Copenhagen: Danish Employers Association.

Chapter 2

Technician Manpower Development in Scotland

Nigel Paine
Scottish Council for Education Technology,
Glasgow, Scotland

INTRODUCTION

Scotland is one of the most northerly countries in Europe. The main landmass consists of a depopulated Highland area that is mountainous and relatively inaccessible, a highly populated urban conglomeration in the center, and a low-population southerly area that marks the border with England. In addition to the mainland, Scotland has approximately 75 inhabited islands stretching off the mainland coast to the northeast. The country's population is approximately 5.2 million people (almost identical to Denmark's), the vast majority of whom (70 percent) live in the greater Glasgow area in the west of Scotland and the Edinburgh urban belt in the east.

The average population density of the districts forming the central belt is about 2.8 persons per ha. (hectare, or 2.47 acres), whereas the overall figure for Scotland as a whole is 0.63 persons per ha. In the Highland area, which occupies 40 percent of the land mass, it is only 0.08 persons per ha.—one of the lowest population densities in the whole of Europe.

Glasgow was once known as the second city of the empire. A heavily industrialized city in the 19th century, growing out of a very successful mercantile city in the 18th, it was fabulously wealthy. Scotland as a whole has been divided by its intensive manufacturing and industrial work in the central belt (steel, coal, shipyards, steam trains, sewing machines, etc.) and its rural, agricultural population in the north and south. This industrial base was transformed in the period after the Second World War and has largely disappeared. The shipyards of the river Clyde, which flows through Glasgow, produced 70 percent of the world's large ships in the 1930s. The shipyards now produce mainly warships under government contract and North Sea oil rigs. The transformation from heavy industrial to sunrise economy has largely been dominated by the successful introduction of electronics industries, which initially chose Scotland as an assembly point in Europe, but which have lately used it as a development and manufacturing center. The companies using Scotland as their major European base include IBM, Compaq, NEC, Motorola, Chungwa, Hughes Electronics, Digital, and many others.

This shift has created a huge expansion in technician-level jobs and, therefore, in the training provision needed to sustain that environment. One of Scotland's recent successes was to persuade Cadence Technology to establish a major research and development base in Livingstone, which lies midway between Glasgow and Edinburgh, as its second base after Silicon Valley. This will provide high-wage, high-skilled jobs building the next generation of microchips.

HISTORICAL OVERVIEW

Under the terms of the Act of Union of 1707, separate governments of Scotland and England united under one parliament (a century after James VI of Scotland united the two kingdoms in 1603 on his accession to the throne of England) to form Great Britain. Scotland's separate legal system, its national church, and the right to issue bank notes were enshrined in the Act of Union. The secretary of state for Scotland ministerial post was established in 1885 at cabinet level to oversee the government of Scotland.

The Scottish Office took over from the Home Office responsibility for the various Scottish boards and for law and order in Scotland. The secretary of state for Scotland also assumed responsibility for the Scotch Education Department, which was renamed the Scottish

Education Department in 1918, the Scottish Office Education Department in 1991, and the Scottish Office Education and Industry Department in 1995.

The secretary of state for Scotland is responsible to Parliament for the work of the Scottish departments, whose executive actions are thus ultimately subject to parliamentary control. As of 2000, Scotland again will have its own parliament, sitting in Edinburgh, with 126 members. Certain areas, including education, will be taken under the full control of the Scottish Parliament. Currently, there are five ministers in the Scottish Office, with a minister of education directly responsible to the secretary of state for the education system in Scotland.

The functions of local government are the responsibility of 32 single-tier councils. These councils serve areas of widely ranging populations (from the Orkney Islands, with a population of 19,760, to the city of Glasgow, with a population of 623,850) and geographic areas (from the city of Dundee, which covers 6,515 ha., to the Highlands, which cover 2,578,379 ha.). Local authorities are directly responsible for school education from preschool to 17 or 18 years of age.

In Scotland, postsecondary education is covered by 43 further education colleges (roughly equivalent to American community colleges). These are directly funded by the Scottish Office with a budget of some £300 million ($500 million) per year. As of April 1999, a further education funding council will be established in Scotland under the administrative aegis of the Scottish Higher Education Funding Council, removing funding from direct government control.

The gross domestic product (GDP) of Scotland totaled £50.7 billion in 1995, which is an average of £9,872 ($16,000) per person. The largest elements in the GDP are financial and business services (20 percent); manufacturing (18 percent); education, social work, and health service (14 percent); and distribution, hotels, and catering (14 percent). The manufacturing sector in Scotland is strongly oriented toward export and 64 percent of Scottish exports go to the European Community. GDP has grown roughly in parallel with the rest of the United Kingdom. On a measure of 1990 = 100, GDP was

89.3 in 1985 and 106.7 in 1995. Roughly 5 percent of GDP is spent on education.

Education in Scotland has a long and distinguished history. By the end of the 15th century, Scotland had three universities, compared to England's two. In the 17th century, the Scottish Parliament passed several acts encouraging the establishment of schools. The Education Act (Scotland), published in 1696, was the world's first national education act, providing for a school and a teacher's salary in every parish. Scotland established a single, external examination system as far back as 1885, and to this day, the Scottish Qualification Authority (SQA) is the single body covering both school and vocational education and training in Scotland.

SCOTLAND'S TECHNICAL EDUCATION PROGRAM

Mission

Scotland prides itself on its education system. Education in Scotland has always enjoyed a high status and most of its key values extend back a number of centuries. Foremost is the provision of free compulsory education from age 5 to age 16. There is also a broadly based curriculum at school level that allows young people to progress into vocational training or higher education, while giving them the broadest choice of continuing occupation. Scotland now has enshrined those principles into a radical reform of the post-16 curriculum, which brings together the academic and vocational parts of the curriculum into a unified program called Higher Still. The first young people will go through this program beginning in August 1999.

Increasingly, Scotland requires an educated population to take control of its own learning destiny. Therefore, the basic tools for lifelong learning need to be built at school and in the immediate postsecondary education framework. A measure of Scotland's success in developing its education system will be its ability to generate lifelong learning habits as well as lifelong learning opportunities throughout the population.

The goal of Scotland's vocational education system is to offer a wide range of programs that lead directly to vocational qualifications or to further education and training. Options include a range of non-advanced vocational and general, pre-employment, and off-the-job training programs, and advanced education leading to higher national certificates (HNC), Higher National Diploma (HND), or degree-level studies. Table 2.1 *(page 40)* shows an outline of the relationships among education partners.

Further Education Colleges

Vocational education is largely provided by the network of 43 further education colleges (FE), which operate from as far south as Dumfries and Galloway (10 or 15 miles from the border with England), to the northernmost tip of Scotland (Thurso College), and in most population points in between. Each college is self-governing and funded directly by a grant from central government through a board of management that comprises up to 16 members. Under the terms of the Further and Higher Education (Scotland) Act of 1992, half the members of each board of management are required to be drawn from local industry and commerce. Of the remainder, four "interested persons" are drawn from the local community, together with the college principal, two members of staff, and a student representative. The board has full executive powers to run the college and reports to the secretary of state through the Scottish Office Education and Industry Department in relation to development planning. The Scottish Office Education and Industry Department makes decisions on funding and takes steps to ensure the quality of outcomes.

Other Key Players

The Scottish Qualifications Authority (SQA) is a statutory body that has responsibility for all qualifications at school and for further education below degree level. SQA also approves education and training

Table 2.1 Relationship among Education Partners in Scotland

Entity	Activity	Roles	Relationship with College
The enabling legislation	Sets framework/ establishes broad policy	Sets general parameters	Fund college activity
Scottish Office Education and Industry Dept. (SOEID)	Implements legislation	Sets general targets	Monitors quality, controls, and supervises
Scottish Qualifications Authority (SQA)	Sets qualification	Controls quality	Establishes curriculum
Local Enterprise Companies (LEC)	Buy services	Ensure training provision	Serve as customer
Local Authority	Influencer	Sit on board	Serve as customer
Local Industry	Monitors	Advises, guides	Serve as customer, adviser
Association of Scottish Colleges (ASC)	Membership association	Influence government, represent views of college	Spokes-person, focal point

establishments that offer courses leading to its qualifications. Much of the curriculum development undertaken by SQA is done in association with experts drawn from colleges and from industry. There is, therefore, an agreed-upon curriculum in Scotland that all providing bodies work with.

Local Enterprise Companies

Beginning April 1, 1994, training in Scotland became the responsibility of the secretary of state for Scotland. A network of local enterprise companies was established under the responsibility of Scottish Enterprise in the lowlands and Highlands and Islands Enterprise in the Highlands. These organizations carry the training budget for Scotland and buy training services from the education system. Much of the specific vocational training for 16-to-18-year-olds is funded in this way.

Data from 1995–96 showed 311,800 students in vocational and further education. This is a dramatic increase from the 1992–93 figure of 207,055 and illustrates the huge increase in participation in further education since the early 1990s. Participation has now stabilized. An additional 25,300 students participate in higher education courses (higher national certificates and higher national diplomas) within the FE system. Increasing numbers of these students are part-time and over the age of 21. A majority of FE students (53 percent) are defined as mature, or over the age of 25.

CURRICULUM

The nonadvanced curriculum (below the first year at university, or skill-based training up to school-leaving level) in further education colleges derives, to a very great extent, from the report *16–18s in Scotland: An Action Plan* (Scottish Education Department 1983). This radical proposal rationalized nonadvanced further education provision and attempted to remove redundancy from the curriculum by creating a modular framework. The outcome of this was a catalog of almost 4,000 national certificate modules. These modules normally last 40 hours each and include time for assessment and reteaching. Each module has a descriptor that specifies the level of entry, the learning outcomes, and the assessment methods. Full-time students normally take 20 modules in a year. Day-release students take 5 or 6 modules a year.

The National Certificate Programme is flexible and updatable, and it offers continuous assessment as its mode of assessment. It is therefore competence based rather than norm referenced. Table 2.1 *(page 40)* lists module groupings.

Because of the modular nature of the certificates, teaching is more flexible. Increasing numbers of modules are available through open (asynchronous) learning either as print materials or as electronic and online materials.

At the advanced further-education level, the Scottish Qualification Authority has modularized offerings into higher units, each of which includes approximately 60 hours of learning and teaching. There are currently 1,200 units in the higher national catalog. See Appendix 2A *(page 52)* for a representative higher national unit.

In addition to national certificates, higher national certificates, and higher national diplomas, the Scottish Vocational Qualification (SVQ) has been phased in at five levels. There are now more than 500 SVQs, which are designed for the workplace by industry-led bodies and relate directly to an individual's ability to do the job. SVQs are based on actual working practices in real workplace conditions and are entirely competency-driven. The SVQ is analogous to the National Vocational Qualifications (NVQ) operating in the rest of the United Kingdom. Both have been recognized as valid qualifications across the European Union.

Within the SVQ framework, a more broadly based prevocational qualification known as the General Scottish Vocational Qualification (gSVQ) has been introduced. This is aimed specifically at 16-to-19-year-olds at school and in further education, as well as at adult returners. The qualification embraces a range of core skills and is designed to provide broad training for employment as well as preparation for progression into higher education. It can also form the precursor to a specific SVQ, but will prepare a young person for a specific job. See Appendix 2B *(page 60)* for a description of a relevant gSVQ.

Table 2.2 Scotland's National Certificate Programme Module Groupings

- Business and Management
- Law, Politics, and Economics
- Arts, Crafts, and Hobbies
- Culture, Society, and Education
- Language, Communications, and Self-Help
- Music and Performing Arts
- Sports, Games, and Recreation
- Food, Leisure, and Tourism
- Environment, Security, Health, and Safety
- Agriculture, Horticulture, and Animal Care
- Sciences and Mathematics
- Health and Personal Care
- Architecture and Construction
- Computers, Electrical, and Electronic Engineering
- Engineering Production and Industrial Design
- Minerals, Materials, and Fabrics
- Transport Services and Vehicle Engineering

TECHNICIAN EDUCATION

The best way to illustrate the range of technician education in Scotland is to trace the vocational qualifications offered, starting with prevocational work and training preparation through national certificate nonadvanced modules into General Scottish Vocational Qualifications. There are also a number of prevocational training programs aimed primarily at the unemployed. The information included concentrates on the Workstart Programme. See Appendix 2C *(page 64).*

Workstart Programme

The Workstart Programme requires six module credits. There are three compulsory credits: Communication for Life and Work; Numeracy for Life and Work; Personal Profiling for Life and Work. Two credits are taken from the following generic modules:

- Skillstart Investigation: The World of Work
- Workstart: Sampling Work
- Skillstart Investigation: Life and Work in a European Country
- Making Local Journeys
- Skillstart Enterprise Activity
- Using a Microcomputer

The final credit must be in a specific skill area, such as catering, construction, office, service, gardening, or mechanical skills.

General Scottish Vocational Qualifications

At the next level are General Scottish Vocational Qualifications. These are generally taken at school or at a further-education college; there are approximately 10 separate gSVQs operating at three levels. They are usually taken over one year and are full-time equivalent study. Level III gSVQ in electronic and electrical engineering is presented in Appendix 2B *(page 60)*.

There are 20 credits in total, 10 of which are compulsory. They cover communication; numeracy; information technology; local investigations; structures and materials; applied electronics; control systems; health and safety in the workplace; either design and graphics in industry or introduction to graphical communication; introduction to quality assurance; and either mathematics, craft technology, analysis and algebra, or core mathematics.

There are then seven credits taken from optional modules such as Power Electronics, Electrical Power Systems, Basic Electrical Plan Safety and Maintenance, Computer System Networks, and Single Phase AC. A three-credit additional assessment that includes problem-solving skills is required.

National Certificates

Of the 4,000 national certificates, 50 or 60 cover technician areas. The national certificate modules are basically the building blocks for Workstart, Skillstart, gSVQs and Scottish Vocational Qualifications. One module, "Introduction to Semiconductor Applications," illustrates the model (see Appendix 2D, *page 66*). Each module covers 40 hours of learning, teaching, and assessment, and has a number of outcomes. In this sample case four outcomes are identified. Each outcome has a number of performance criteria. To satisfactorily achieve the outcome, the student must attain all of the performance criteria. There are 13 performance criteria associated with this module. Modules also contain a content and context statement and some suggested learning and teaching approaches.

Advanced-level technician education is covered by higher national units, which can be combined to form a higher national certificate, equivalent to one year full time, or a higher national diploma, equivalent to two years full time and up to and including the first year of a degree program.

Higher National Units

There are nearly 1,200 higher national units, each of which covers 60 hours of learning. Appendix 2A *(page 52)* presents information on the Electronic Component Technology and Design unit. The unit structure is relatively similar to the national certificate structure. There is a description of the unit, the outcomes, and the credit value. In this case five outcomes are identified. Each outcome has a number of performance criteria and a range statement defining the context in which the performance criteria are judged. For example, the performance criterion states, "The visual identification of different types of electronic components is correct." The range statement will define electronic components as "passive, active, discreet, and integrated." There is then a statement of how the evidence should be gathered to demonstrate competence and, finally, a statement on assessment.

Advantages of the Scottish Framework

The Scottish system has a number of advantages, including the following:

1. Flexibility. Programs of study can be put together almost on demand, based on modules and units. It is also easy to define national qualification through amending specific modules or units or by revising what modules are relevant. This is undertaken through a network of sector groups, usually chaired by an industrialist or someone with industry experience who oversees the quality and relevance of the module and unit framework.

2. Multipurpose. The units and modules can be used in different contexts. For example, students can fill a timetable by taking one or two national certificates in their senior year. On the other hand, in the same school, students following a vocational program may take an entire national certificate program, either as a General Scottish Vocational Qualification or as another kind of prevocational qualification. Equally, students following higher-education courses can "in-fill" with modules or units. For example, in Information Technology, students can broaden their skill base by adding modules or units to their learning program.

3. Transparency. The entire curriculum is documented. The national certificate catalog and the higher units catalog are both produced on CD-ROM by the Scottish Council for Education Technology (SCET) and distributed freely in colleges, schools, training establishments, and companies. A training manager is able to read the basic course structure and work out how closely it meets the needs of trainees.

4. Multicourse use. The same national certificate module can be used in a tailored award, a national certificate full-time course, a vocational qualification, or curriculum in-fill. Because the building blocks of the curriculum are relatively small, the curriculum can be redefined quickly and simply.

5. Quality assurance. Because all programs are taught within a similar competence framework using the same model of performance criteria, range statements, and context, both internal quality assurance by the colleges and external moderation are more easily undertaken. SQA's policy is to drive the quality assurance function down to the colleges themselves. Colleges are put through rigorous training and testing procedures to maintain this role.

6. Meeting industry needs. If the needs of industry change, a new module or unit can be put together very quickly. Some of these units are "commissioned" by SQA, but others can be developed by colleges and companies working together. If a module or unit does not exist, it can be written and put into the catalog after SQA validation. Catalogs are updated every year.

7. Transferability. The transparency of the curriculum also allows credits to be transferred between institutions. A student who achieves a number of higher national certificates from one college or obtains higher national units from one college will usually be able to transfer these into another course at another institution. While this practice is very common in America, it is relatively rare in the United Kingdom.

INVESTORS IN PEOPLE

There are strong pressures from government and Scottish enterprise to encourage companies to obtain Investors in People (IP) status. This is an externally audited award given to companies or organizations that are deemed to have met these four standards:

Commitment. An Investor in People makes a public commitment from the top to develop all employees to achieve its business objectives.

Review. An Investor in People regularly reviews the training and development needs of all employees.

Action. An Investor in People takes action to train and develop individuals upon recruitment and throughout their employment.

Evaluation. An Investor in People evaluates the investment in training and development to assess achievement and improve future effectivencss.

Currently, 883 companies in Scotland have achieved IP status and another 3,278 companies are committed to achieving it. In short, all IP-status companies guarantee their employees a personal development program, an overall company training program, and rigorous attention to employee training and development within the context of business success.

LECTURER TRAINING AND CONTINUING EDUCATION

Scotland has a Scottish School of Further Education. This is part of the Faculty of Education of Strathclyde University, which awards a Teaching Certificate in Further Education (TQFE) to further-education lecturers transferred temporarily from their posts to undertake this certificate on a part-time and distance-learning basis. The TQFE is not compulsory, but all lecturers are given the opportunity to obtain a TQFE as soon as they enter a college.

The course consists of two four-week periods of release time and two one-week placements for release time spread over two years. Between the blocks of release time, the course is taught by distance learning and independent study. In addition, there are specific training courses. These are currently organized by SCET in the area of learning technology and by the Scottish Further Education Unit in other curriculum areas.

Many colleges organize their own in-service training or combine to offer in-service training. Every college in Scotland is required to have a staff development officer who oversees a program of both curriculum and skill development. Many colleges have obtained or are in the process of obtaining IP certification; therefore, they are committed to providing personal development plans and training schemes for all employees, not just academic staff.

PROGRESSION

In an attempt to integrate Scottish Vocation Qualifications, progression has been built in to take someone from, say, a gSVQ through to higher education qualifications and SVQs. Appendix 2E *(page 71)* illustrates an electrical technician coming in at gSVQ III, leading to HNC qualification, then to HNDs or an SVQ. Such progression models are available throughout Engineering and Manufacturing Technology, Business Administration and Management, Hospitality Travel and Tourism, Communications and Media, Arts and Social Sciences, Built Environment (construction, building design, architecture), Hairdressing and Beauty Services, and other specialties.

CONCLUSION

The Scottish vocational training system has been highly successful in its modified form since the early 1980s. Throughput of students has increased by over 40 percent during that period and the flexibility and modularity of the system have been widely praised in Europe.

The recent merger of the former Scottish Vocational Qualification Council and the Scottish Examination Board to form the Scottish Qualifications Authority harmonizes school and vocational qualifications and paves the way for the delivery of the new Higher Still qualification running through from vocational to academic. This should set new standards in coherent delivery. Scotland is confident that it can meet the skill needs of employers and the learning expectations of its population using the framework that has been established.

The government is keen to increase learning and has recently published a discussion paper, *The Learning Age* (Department for Education and Employment 1998), to raise key issues around the concept of learning technology. There are also plans to launch a University for Industry (UFI) in the year 2000, which will market the concept of learning and offer guidance and information on provision and qualifications. This UFI initiative will have a distinctive Scottish flavor. As a membership organization, it will work with employers, students, and workers to dramatically increase access to and provide a range of learning opportunities. In Scotland, the first approach will encourage skill development in the electronic industry at the technician level.

There have been huge changes in vocational education and training since the early 1980s, and the changes will continue into the next century. Scotland cannot afford to lose the opportunities that the new global economy offers it. Many challenges still lie ahead.

ABOUT THE AUTHOR

Nigel Paine has been chief executive of the Scottish Council for Educational Technology (SCET) in Glasgow, Scotland, since April 1990. SCET is a company that promotes the effective use of new learning technologies in education and training. Its mission is "to transform learning throughout life by harnessing the power of technology." Paine is a fellow of the Institute of Personnel Development and was recently elected a fellow of the Royal Society of Arts. He is a visiting professor at Napier University in Edinburgh.

BIBLIOGRAPHY FOR CHAPTER 2

The British Council. 1993. *Education in Scotland.* Edinburgh, Scotland: The British Council.

Department for Education and Employment. 1998. *The Learning Age.* London: Department for Education and Employment.

Industrial Research and Development Advisory Committee of the European Commission. 1994. *Quality and Relevance: The Challenge of European Education.* Brussels: Industrial Research and Development Advisory Committee of the European Commission.

Louden, Dick. 1997. *The Story of SCOTVEC 1985–97.* Glasgow, Scotland: Scottish Vocational Education Council.

Scottish Education Department. 1983. *16–18s in Scotland: An Action Plan.* Edinburgh, Scotland: Scottish Education Department.

Scottish Enterprise. 1996. *Scottish Enterprise: Developing a Learning Technology Strategy for the Scottish Enterprise Network: Consultative Paper.* Edinburgh, Scotland: Scottish Enterprise.

———. 1997. *Scottish Enterprise Annual Report 1996/7: Environment, Entrepreneurs, Jobs, Skills, Exports, People, Inward Investment.* Glasgow, Scotland: Scottish Enterprise.

Scottish Office. 1997. *Raising the Standard: A White Paper on Education and Skills Development in Scotland.* Edinburgh, Scotland: Scottish Office.

Scottish Office Education and Industry Department. 1983. *Scottish Education Department: The Action Plan.* Edinburgh, Scotland: Scottish Office Education and Industry Department.

———. 1997. *Scottish Office Education and Industry Department: Education and Training in Scotland—A National Dossier.* Edinburgh, Scotland: Scottish Office Education and Industry Department.

Scottish Qualifications Authority. 1997. *SQA Annual Review 1996/7.* Glasgow, Scotland: Scottish Qualifications Authority.

Scottish Vocational Education Council. 1992. *SCOTVEC, Highlands and Island Enterprise, Scottish Enterprise, Scottish Quality Management System 1992.* Glasgow, Scotland: Scottish Vocational Education Council.

———. 1994. *SCOTVEC Setting Standards for Scotland.* Glasgow, Scotland: Scottish Vocational Education Council.

Appendix 2A: Scottish Qualifications Authority (SQA) Higher National Unit Specification

GENERAL INFORMATION

Unit Number	2451445	Cognate Group 422
Superclass	XL	
Title	ELECTRONIC COMPONENT TECHNOLOGY AND DESIGN	

DESCRIPTION

GENERAL COMPETENCE FOR UNIT: Demonstrating an ability to identify components, to understand potential failure mechanisms and the use of polymeric materials, and to appreciate relevant selection criteria and the driving forces behind emerging technologies.

OUTCOMES

1. Differentiate between electronic components;
2. Describe typical failure mechanisms for common electronic components and appropriate methods of handling and storage;
3. Explain the uses, properties and methods of application of key polymeric materials used in advanced electronics assembly;
4. Identify selection criteria for printed wiring board (PWB);
5. Outline the trends in emerging technologies.

CREDIT VALUE: 1 HN Credit

ACCESS STATEMENT: Access to this unit is at the discretion of the centre. However, it would be beneficial if the candidate had familiarity with electronic components and design concepts.

For further information contact: Committee and Administration Unit, Scottish Qualifications Authority, Hanover House, 24 Douglas Street, Glasgow G2 7NQ.

HIGHER NATIONAL UNIT SPECIFICATION
STATEMENT OF STANDARDS

UNIT NUMBER: 2451445

UNIT TITLE: ELECTRONIC COMPONENT TECHNOLOGY AND DESIGN

Acceptable performance in this unit will be the satisfactory achievement of the standards set out in this part of the specification. All sections of the statement of standards are mandatory and cannot be altered without reference to SQA.

OUTCOME

1. DIFFERENTIATE BETWEEN ELECTRONIC COMPONENTS

PERFORMANCE CRITERIA

(a) Visual identification of different types of electronic components is correct.
(b) Determination of electronic component values by electrical test is correct.
(c) Identification of different electronic component package types is clear and precise.

RANGE STATEMENT

Electronic components: passive; active; discrete; integrated.

Electrical test: resistance; capacitance; inductance.

Package types: axial; radial; chip; Metallised Electrode Leadless Face bonding (MELF) devices; single-in-line packages; dual-in-line packages; small outline transistors and integrated circuits; quad flat packs with J-lead and gull-wing leads.

EVIDENCE REQUIREMENTS

Written or performance evidence of the candidate's ability to describe the components, marked component values, and package types on a complex printed circuit assembly.

Performance evidence of the candidate's ability to identify the values of, and differentiate between, visually similar components using standard bench-test equipment.

OUTCOME

2. DESCRIBE TYPICAL FAILURE MECHANISMS FOR COMMON ELECTRONIC COMPONENTS AND APPROPRIATE METHODS OF HANDLING AND STORAGE

PERFORMANCE CRITERIA

(a) Explanation of failure mechanisms is clear and precise.
(b) The main techniques in the control of electrostatic discharge (ESD) are explained correctly.
(c) The reasons for desiccant packaging and for limitations on component shelf life are clearly explained.

RANGE STATEMENT

Failure mechanisms: electrical—ESD, overvoltage, reverse polarity; mechanical damage—cracks, chips, bent leads, coplanarity; environmental—contamination, delamination, water ingress.

EVIDENCE REQUIREMENTS

Written or oral evidence of the candidate's ability to describe failure mechanisms.

Written or oral evidence of the candidate's ability to explain the control of ESD.

Written or oral evidence of the candidate's ability to describe the reasons for desiccant packaging and limitations on component shelf life.

OUTCOME

3. EXPLAIN THE USES, PROPERTIES AND METHODS OF APPLICATION OF KEY POLYMERIC MATERIALS USED IN ADVANCED ELECTRONICS ASSEMBLY

PERFORMANCE CRITERIA

(a) The uses of polymeric materials within electronics are identified correctly.
(b) The properties of key polymeric materials used are explained correctly.
(c) Appropriate methods of application for these materials are explained correctly.
(d) Relevant environmental and health and safety issues are correctly identified.

RANGE STATEMENT

Uses: component manufacture; printed wiring board; assembly.

Properties: mechanical strength; adhesion; flammability rating; temperature coefficient of expansion.

Application methods: printing; pin transfer; dispensing.

EVIDENCE REQUIREMENTS

Written or oral evidence of the candidate's ability to explain the properties of key resin materials used in advanced electronics assembly and appropriate methods of application.

Written or oral evidence of the candidate's ability to identify relevant environmental and health and safety issues.

OUTCOME

4. IDENTIFY SELECTION CRITERIA FOR PRINTED WIRING BOARD (PWB)

PERFORMANCE CRITERIA

(a) Different types of printed wiring board are clearly described.
(b) The criteria for component selection applicable to different types of assembly are clearly explained.
(c) The properties which determine constraints on design and assembly are clearly explained.

RANGE STATEMENT

PWB types: single layer; plated through-hole (PTH); multi-layer.

Component selection criteria: package form; complexity/pin count; power rating; mechanical components.

Properties: temperature range; humidity rating; power dissipation; electrical characteristics.

EVIDENCE REQUIREMENTS

Written and graphical evidence of the candidate's ability to describe the differences between single layer, PTH, and multi-layer PWBs, the criteria for choice of components appropriate to each type, and the physical properties of PWBs which determine constraints on use.

OUTCOME

5. OUTLINE THE TRENDS IN EMERGING TECHNOLOGIES

PERFORMANCE CRITERIA

(a) Constraints and limitations of current component technologies and manufacturing processes are clearly specified.
(b) Alternatives to provide solutions to the above constraints are clearly described.

RANGE STATEMENT

Component constraints and limitations: lead pitch; component density; propagation delay.

Manufacturing process constraints and limitations: quality; yield; cost.

EVIDENCE REQUIREMENTS

Written evidence of the candidate's understanding of the constraints and limitations of current technology, and of alternative solutions designed to overcome these.

MERIT: Pass with merit may be awarded to a candidate who achieves the performance criteria for all outcomes and in doing consistently demonstrates superior performance, for example by

(a) displaying a greater depth of knowledge of the operation, selection for use and failure mechanisms associated with components;
(b) displaying evidence of further reading or other additional effort.

ASSESSMENT

In order to achieve this unit, candidates are required to present sufficient evidence that they have met all the performance criteria for each outcome within the range specified. Details of these requirements are given for each outcome. The assessment instruments used should follow the general guidance offered by the SQA assessment model and an integrative approach to assessment is encouraged. (See references at the end of support notes.)

Accurate records should be made of the assessment instruments used showing how evidence is generated for each outcome and giving marking schemes and/or checklists, etc. Records of candidates' achievements should be kept. These records will be available for external verification.

SPECIAL NEEDS

Proposals to modify outcomes, range statements or agreed assessment arrangements should be discussed in the first place with the external verifier.

UNIT NUMBER: 2451445

UNIT TITLE: ELECTRONIC COMPONENT TECHNOLOGY AND DESIGN

SUPPORT NOTES: This part of the unit specification is offered as guidance. None of the sections of the support notes is mandatory.

NOTIONAL DESIGN LENGTH: SQA allocates a notional design length to a unit on the basis of time estimated for achievement of the stated standards by a candidate whose starting point is as described in the access statement. The notional design length for this unit is 40 hours. The use of notional design length for programme design and timetabling is advisory only.

PURPOSE: This unit is designed for manufacturing, maintenance and process technicians in the advanced electronics assembly industry.

CONTENT/CONTEXT: This unit should be delivered in the context of manufacturing equipment which is in use for automated assembly of surface mount and associated components.

In the Range Statement for Outcome 4, the term " mechanical components" refers to elements of the assembly such as connectors and screening cans, where mechanical issues such as the integrity of the assembly are more important than electronic criteria.

There are a number of emerging package styles (such as Tape Automated Bonding, Ball Grid Arrays, Multi-Chip Modules), and it is recommended that these should be covered as part of Outcome 5 rather than Outcome 1, with the implications for assembly covered within HN Unit: 2451435 Component Placement which would be presented concurrently.

A number of topics have been identified which should be highlighted throughout the course even where they are not explicit in the outcomes assessed:

> the critical importance of PPM (parts-per-million) quality, yield, cost and reliability; awareness of health and safety issues, including relevant statutory requirements and accepted Codes of Practice.

Companies offering placements for candidates, shall instruct the candidate in any elements of health and safety, COSHH and related topics which may be necessary for them to work safely in a production environment.

REFERENCES

1. Guide to unit writing.
2. For a fuller discussion on assessment issues, please refer to SQA's Guide to Assessment.
3. Information for centres on SQA's operating procedures is contained in SQA's Guide to Procedures.
4. For details of other SQA publications, please consult SQA's publications list.

Appendix 2B: Specification for General Scottish Vocational Qualification National Certificate (Level III) Engineering—Electronic and Electrical

Award number: GG0404601

Total credit value of award: 20 credits

1 Mandatory Modules

A total of **10 credits** must be gained by achieving the following mandatory modules:

No.	Title	Credit
7110045	Communication 3	1
7180225	Numeracy 3	1
8111025	Information Technology 3	1
7350605	Local Investigations 3*	1
2150186	Structures and Materials 1	0.5
2150206	Applied Electronics 1	1
2150196	Control Systems 1	1
7161324	Health and Safety in the Work Place	1
	1 credit from:	1
8150176	Design and Graphics in Industry	
2281042	Introduction to Graphical Communication	
81162	Introduction to Quality Assurance	0.5
	1 credit from:	1
91045	Mathematics: Craft Technology 1 (0.5)	
91046	Mathematics: Craft Technology 2 (0.5)	
7180401	Mathematics: Analysis/Algebra 1	
7180331	Core Mathematics 4	

Total credit	10

* Personal and interpersonal skills at stage 3 are built into this module.

2 Optional Modules

A total of **7 credits** must be gained from the following optional modules:

No.	Title	Credit
64163	Switchgear and Protection	1
64165	Power Electronics	1
64203	Communication Radio Circuits and Systems 1	1
64205	Telecommunications Lines: Cables	1
2150096	Fault Diagnosis on Basic Electronic Circuits	1
2150136	Fault Diagnosis on Complex Electronic Circuits and Systems	1
71119	Computer System Software	1
2180006	Electrical Power Systems	1
84167	Electrical Machine Principles	0.5
84168	Electrical Motor Applications	0.5
84169	Basic Electrical Plant Safety and Maintenance	1
84170	Electrical Plant Maintenance	1
2150020	Logic Families and Digital Systems Analysis	1
2150030	Sequential Logic	1
2150040	Digital MSI Devices	1
2150050	Microcomputer Hardware	1
2150060	Microprocessor Fault Diagnosis	1
2150070	Peripherals	1
2150080	Computer System Networks	1
2150220	Electronic Components and Circuit Assembly Techniques	1
2150240	Amplification	1
2150146	Frequency Generation and Selection	1
2150260	Integrated Circuit Applications	1
2150270	Power Supplies	1

No.	Title	Credit
2160020	Circuit Elements	0.5
2160030	Single Phase AC	1
2160040	Transformation and Rectification	0.5
2160050	Electromagnetics	1
2160060	Electrostatics	0.5
2160070	Power Factor Improvement and Three-Phase Theory	0.5
2160080	Network Analysis	0.5
2160090	DC and AC Circuit Responses	0.5
3150584	Measurement Technology: Pressure/Level	1
3150594	Measurement Technology: Flow	1
3150604	Measurement Technology: Temperature	1
3150614	Signal Conditioning in Telemetry	1
3150644	Programmable Logic Controllers	1
7180414	Mathematics: Analysis/Algebra 2	1
	0.5 credit or 1 credit from:	
7181144	Mathematics: Calculus 1	
7181155	Mathematics: Calculus A (0.5)	
8112032	Computer Software	1
2150116	Combinational and Sequential Logic	1
2160010	Electrical Fundamentals	1
2150010	Combinational Logic	1
2150230	Introduction to Semiconductor Applications	1
2150024	Wiring and Assembly Techniques	0.5
2150034	Soldering Techniques on Electronic Circuits	1
4110060	Materials and Jointing Methods (Electrical)	0.5
2150410	Introduction to Electronic Test Equipment and Measurements	0.5
64200	Basic Telecommunications	1

No.	Title	Credit
64201	Principles of Telecommunication Systems	1
64209	Telecommunications: Digital Transmission Techniques	1
64204	Telecommunications: Line Transmission	1
3150624	Program Controlled Systems	1
3151044	Complex Program Controlled Systems	1
2260100	Computing in Engineering 1	1
2150101	Introduction to Basic Semi-Conductor Manufacturing Procedures	1
2310011	Basic Micro-Soldering	1

Total credit	7

3 Additional Assessment

A pass in the additional assessment for Engineering: Electronic and Electrical is also mandatory for this award.

Additional assessment	Total credit 3

Problem solving skills at stage 3 are built into the additional assessment.

Total credit of award	7

MMcA
DE36111297.1
9 January 1998

Appendix 2C: Specification for National Certificate—Workstart

Award number: NN1101502

Total credit value of award: 6 credits

1 Mandatory Core Skills Modules

A total of **3 credits** must be gained by achieving the following mandatory core skills modules:

No.	Title	Credit
7110065	Communication for Life and Work	1
7180005	Numeracy for Life and Work	1
7350005	Personal Profiling for Life and Work	1

Total credit	3

2 Generic Modules

A total of **2 credits** must be gained from the following generic modules:

No.	Title	Credit
91090	Using a Microcomputer	1
7197017	Skillstart Enterprise Activity	1
7350773	Making Local Journeys	1
7351607	Skillstart Investigation: The World of Work	1
7350015	Workstart: Sampling Work	1
7350025	Investigating Life and Work in a European Country	1

Total credit	2

3 Optional Activity-Based Modules

A total of **1 credit** must be gained from the following optional modules:

No.	Title	Credit
7190125	Workstart: Catering Skills	1
7190065	Workstart: Construction Skills Using Wood	1
7190075	Workstart: Construction Skills Using Metal	1
7190035	Workstart: Office Skills	1
7190025	Workstart: Service Skills—Caring	1
7190015	Workstart: Service Skills—Retail	1
7190055	Workstart: Gardening Skills	1
7190045	Workstart: Mechanical Skills	1

Total credit	1

Total credit of award	6

MMcA
DE48111297.1
9 January 1998

Appendix 2D: National Certificate Module Descriptor

Module Number	**2150230**	**Session 1990-91**
Superclass	**XL**	
Title	**INTRODUCTION TO SEMICONDUCTOR APPLICATIONS**	

DESCRIPTION

Purpose

This module has been designed for use at craft and technician level by individuals who require an understanding of the operation and application of semiconductor diodes and amplifiers.

Preferred Entry Level

2150220 Electronic Components and Circuit Assembly Techniques.

Outcomes

The student should:

1. interpret the operation of semiconductor diode circuits;
2. outline the use of power control devices;
3. interpret the operating conditions of a single-stage resistance loaded small-signal amplifier;
4. test operational amplifier circuits.

Assessment Procedures

Acceptable performance in the module will be satisfactory achievement of all the Performance Criteria specified for each Outcome.

The following abbreviations are used below:

PC Performance Criteria
IA Instrument of Assessment

Note: The Outcomes and PCs are mandatory and can not be altered. The IA may be altered by arrangement with SQA. (Where a range of performance is indicated, this should be regarded as an extension of the PCs and is therefore mandatory.)

OUTCOME 1 **INTERPRET THE OPERATION OF SEMICONDUCTOR DIODE CIRCUITS**

PCs

(a) The measurement and recording of diode circuit voltage levels are correct.

(b) The identification of operational limitations from manufacturers' data sheets is correct.

(c) The explanation of the circuit operation is correct.

IA Assignment

The student will be set tasks which test the ability to interpret the operation of two different semiconductor diode circuits. The student will be required to take measurements of diode circuit voltage and current levels and answer questions which relate these measurements to manufacturers' data sheets. The measurements will be recorded in a pre-specified format.

The student will also be required to maintain a log book which includes a brief explanation of the operation of the circuit.

Satisfactory achievement of the Outcome will be based on the student attaining all of the Performance Criteria and correctly answering the questions connected with Performance Criteria (b) and (c).

OUTCOME 2 **OUTLINE THE USE OF POWER CONTROL DEVICES**

PCs

(a) Power control devices are correctly identified from their symbols.

(b) The function of power control devices is correctly stated.

(c) Typical applications for power control devices are correctly stated.

IA Structured Question

The student will be set a structured question to test understanding of the use of power control devices.

The student will be given a circuit diagram(s) containing a thyristor, diac and triac which he/she will be required to identify from their symbols. The student will be required to state the function and a typical application for each.

Satisfactory achievement of the Outcome is based on all parts of the question being correctly answered.

OUTCOME 3 **INTERPRET THE OPERATING CONDITIONS OF A SINGLE-STAGE RESISTANCE LOADED SMALL-SIGNAL AMPLIFIER**

PCs

(a) The measurement and recording of circuit voltage levels are correct.

(b) The measured voltages are correctly related to the biasing of the transistor.

(c) The gain of the circuit is correctly determined from measurement of input and output signal amplifiers.

(d) The explanation of the circuit operation is correct.

IA Assignment

The student will be set a task which tests the ability to interpret the operating conditions of a single-stage, resistance loaded small-signal amplifier. The student will be required to take measurements of circuit voltage levels and relate these measurements to the biasing of the transistor and gain of the amplifier. The measurements will be recorded in a pre-specified format.

The student will also be required to maintain a log book which includes a brief explanation of the operation of the circuit.

Satisfactory achievement of the Outcome will be based on the student attaining all of the Performance Criteria and correctly answering the questions connected with Performance Criteria (b) (c) and (d).

OUTCOME 4 **TEST OPERATIONAL AMPLIFIER CIRCUITS**

PCs

(a) Appropriate adjustments are made to the circuit to provide a voltage null at the output.

(b) The gain of inverting and non-inverting configurations is correctly calculated from measurements taken and recorded.

(c) The phase relationship between input and output signals in inverting and non-inverting configurations is correctly measured and recorded.

IA Practical Exercise

The student will complete practical exercises to demonstrate an ability to test operational amplifier circuits.

The student will be given two preconstructed units in which adjustments and measurements are to be made and recorded.

The exercise will be carried out in conjunction with a suitably constructed observation checklist.

Satisfactory achievement of the Outcome will be based on all Performance Criteria being met.

The following sections of the descriptor are offered as guidance. They are not mandatory.

CONTENT/CONTEXT

Safety regulation and safe working practices should be observed at all times.

Corresponding to Outcomes 1-4:

1. Signal diodes, rectifiers and zener diodes. Manufacturers' data for peak inverse voltage, maximum forward current and typical forward voltage drop with reference to device characteristics. Testing of diode and labelling of terminals using digital or multimeter.

   Diode applications to include rectification, clipping, clamping, and voltage stabilisation.

2. Thyristor, diac, triac.

3. Only common-emitter and common source configuration to be investigated.
4. Only inverting and non-inverting configurations to be investigated.

   Non-inverting configuration to include voltage-follower.

SUGGESTED LEARNING AND TEACHING APPROACHES

This module must be taught in a workshop environment, since it involves the operating conditions of devices in practical circuits.

In all Outcomes, pre-constructed circuits are used, such that the student has access for adjustment and measurement.

Appendix 2E: Progress of Electrical Technician

Electrical Technician

(Electrical Engineering)

Supervises craft electricians carrying out installation, testing, and machine maintenance of electrical plant and equipment.

Progression

To pursue a career such as Craft Electrician, **you need** GSVQ III Engineering: Electronic and Electrical Engineering

To pursue a career such as **Electrical Technician**, you need HNC Electrical Power Engineering, or HNC Electronic and Electrical Engineering ☞

To pursue a career such as Incorporated Engineer, **you need**
HND Electronic and Electrical Engineering, or
HND Electronic and Electrical Engineering with Management

Related Qualifications
- HNC Electrical Engineering
- HNC Electrical Engineering with Electrical Plant
- HNC Electronic and Electrical Engineering

Higher Qualifications
- HND Electronic and Electrical Engineering

SVQs
- SVQ III

Chapter 3

Effective Manpower Development Systems in Australia

Peter Noonan
Australian National Training Authority,
Brisbane, Queensland

INTRODUCTION

Australia's vocational education and training system has grown to equip Australians with the skills and learning required by enterprises and industry. Australia's manpower development system—known as Vocational Education and Training (VET)—began with the founding of the first technical college in Hobart in 1827. For many years in Australia's federal system, VET was confined to a narrow range of qualifications offered by state-based technical and further education (TAFE) institutes that were managed from state education departments. It was not until 1974, when the federal government commissioned a report on the sector, that TAFE was put on the education map.

The *Kangan Report* (named after the chair of the committee) articulated a role for TAFE to provide a significant educational experience for individuals within the context of supporting industry requirements. The report noted that:

> The emphasis in technical college type institutions should be primarily on the needs of the individual for vocationally oriented education and the manpower needs of industry should be the context of these courses (Australian Committee on Technical and Further Education 1974).

The report also saw TAFE as a significant means of giving people opportunities to facilitate lifelong learning, noting that "Ideally, through recurrent educational opportunities, people who so wish should be able to repair, as often as necessary, insufficiencies of their initial education. They should also be able to add to, or replenish, their education."

Since then, TAFE has given millions of Australians the skills to lead a productive working life. It has been instrumental in developing the skills pool upon which Australian industry so heavily depends, and it has given many of its students a chance to acquire worthwhile educational qualifications.

In the years since the *Kangan Report*, the TAFE sector has gained an enhanced reputation by taking national approaches to curriculum and qualifications. During this time, TAFE grew in public esteem as a significant pathway to employment and workplace advancement. As a consequence, the demand for TAFE has expanded considerably. In 1974, the *Kangan Report* noted that TAFE enrollments were at 349,000 (excluding adult education). In 1996, the number of publicly funded TAFE enrollments for vocational programs was about 1.1 million. Currently, there are nearly twice as many TAFE students as university students.

A single qualification framework—the Australian Qualifications Framework (AQF)—spans the schools, VET, and higher education sectors. It is a nationally consistent framework providing standards categorized from levels 1 through 12 for all qualifications recognized in postcompulsory education. The benefit of this framework is that it allows for credit transfer and articulation between qualifications. The AQF consists of guidelines that define each qualification along with the principles and protocols covering articulation, issuing of qualifications, and transition arrangements. Schools deliver Certificates of Senior Secondary Education after 12 years of schooling, and Certificates 1 to 4. The VET sector delivers certificates 1 to 4, diplomas, and advanced diplomas. Diplomas through doctoral degree levels are delivered by the higher education sector.

A diverse range of courses leading to VET qualifications are offered at 1,132 locations across the country through TAFE institutes, 514 community centers, and approximately 2,500 registered private training providers (including industry and enterprise providers). More than 1.3 million people annually undertake publicly funded vocational education and training programs. For those who have not successfully completed their schooling, general education remains an important part of vocational education and training provision, particularly in TAFE.

Over the past decade Australian governments and industry have engaged in a major process of reform to vocational education and training to meet challenges proposed since the *Kangan Report*. The forces driving those challenges—globalization, the tailoring of products and services to meet specific client needs, the impact of new technology, changes in work organization, and demographic and social changes—are accelerating. As the trend intensifies for new job growth to be knowledge-based rather than resource-based—and as Australia seeks to improve its international competitiveness—the importance of vocational education and training increases.

All Australians require some postsecondary education and training to avoid disadvantages in the labor market and to ensure that Australia's skills profile does not lag behind its competitors. The "half-life" of qualifications is decreasing as a result of technological and structural change. The demand for vocational education and training places and qualifications will grow in the period ahead. Individuals and enterprises will also demand different sorts of outcomes besides whole qualifications provided in different ways from traditional institutional provider approaches. These challenges mean that Australia's VET system will focus on building a client-focused culture, promoting opportunities for lifelong learning, advancing a national identity for VET, and rewarding innovation and best-practice approaches.

MISSION STATEMENT AND MANPOWER DEVELOPMENT

How the Vocational Education and Training System Works

The national VET system is a cooperative arrangement between the commonwealth, state and territory governments, industry and industry training advisory bodies (ITABs), private and public training providers, and other stakeholders. The VET sector in Australia consists of state and territory public TAFE systems; adult and community education (ACE) institutions that deliver VET; and private providers including schools, community organizations, enterprises, and industry bodies that deliver nationally recognized vocational programs. Cooperation is directed toward achieving a world-class training system responsive to all clients and their training needs.

The Australian National Training Authority (ANTA) has helped to facilitate this national focus for VET. ANTA was established through an agreement of the Australian states and territories in response to the need for a national focus for vocational education and training, with strong industry input. Its mission statement is

> To help Australia become a more internationally competitive and equitable society by building a national vocational education and training system which is responsive to the needs of industry and individuals. This system should deliver world class VET, enhance the employment prospects of Australians and be seen by them as a critical ingredient of their success (ANTA 1998).

ANTA consults with industry and is responsible for development and advice on national policy, objectives of a national strategic plan, and state training profiles for endorsement by the Ministerial Council (MINCO). The authority is also responsible for administering a number of national programs and projects and the commonwealth funds for VET. ANTA is run by an industry-led board that advises and reports to a ministerial council composed of relevant commonwealth, state, and territory ministers.

NATIONAL ADMINISTRATIVE STRUCTURE

Structure and Processes of the National Vocational Education and Training (VET) System

Figure 3.1 shows an outline of the national VET system and its key relationships.

Figure 3.1 Critical Relationships in Australia's Technical Education System

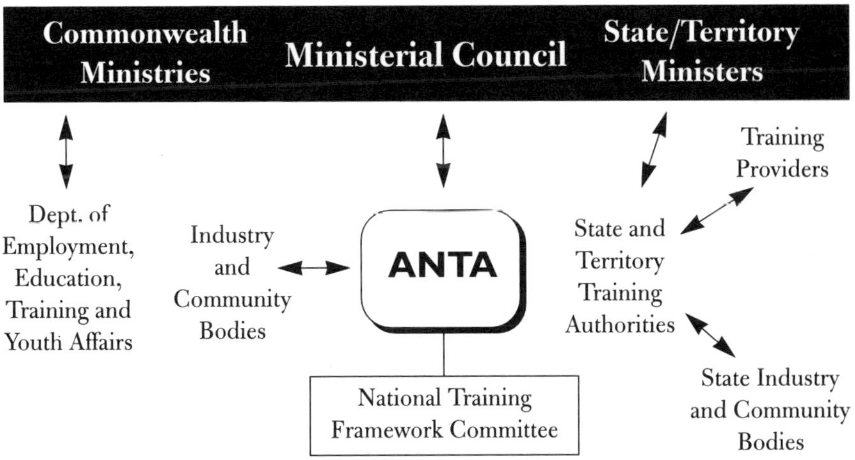

Source: Australian National Training Authority, 1997d

The last 15 years of VET have been a time of great change, and the last five years have intensified this trend. In 1994, ANTA developed the first national strategy for VET—*Towards a Skilled Australia*—which set broad strategic directions for VET over the medium term. It also outlined specific initiatives in relation to responsiveness, quality, accessibility and efficiency to pursue in 1995 and 1996. The strategy provided for a planning framework for use by industry and state and territory training authorities. A second strategy—A Bridge to the Future—was agreed in 1998.

A review of the VET system in 1995 identified areas of significant achievement and areas requiring further improvement. Feedback from industry indicated that the training system was still too complex. There were problems with the national framework for the recognition of training, including lack of coverage of employment-based training contracts, and mechanisms for the provision of advice from industry were not fully effective. In addition, there were problems with the collection of comparable management information statistics on VET. During 1996, there was significant attention given to the following issues:

- reducing the complexity of the training system and the way in which skills are recognized;
- increasing the quality and relevance of employment-based training;
- enabling industry to provide advice on training issues more effectively;
- increasing the comparability of management information statistics across systems to drive system improvements.

The National Training Framework

The purpose of the National Training Framework (NTF) is to make the regulation of Australia's national training arrangements streamlined and more flexible, enabling training to be more responsive to demand. The NTF provides a comprehensive approach to define the relationships between industry bodies, state and territory training authorities, training organizations, and ANTA. The purpose is to produce high-quality training that meets the needs of industry because it is nationally portable, flexible in its delivery, and responsive to client needs.

There are two interconnected features of the NTF: revised and simplified arrangements for the recognition of training organizations that ensure the quality of training provision, called the Australian Recognition Framework (ARF), and skill standards, called training packages, that integrate nationally available training products and include industry or enterprise competency standards. The skill standards are packaged against qualification levels of the AQF and nationally endorsed assessment guidelines. Training packages can also include learning strategies, professional development materials, and assessment materials. The packages are developed through an industry-managed process.

The National Training Framework Committee (NTFC) was formed in 1996, consisting of a small business-led committee. The NTFC is responsible for developing policy to implement the NTF and to endorse the key components of national training packages. The primary role of the committee is to provide advice to the ANTA board on a range of matters critical to the operation of the National Training System, including training packages, assessment policy, national recognition arrangements, qualifications, and data collection. Figure 3.2 identifies key components and relationships of the NTF.

Figure 3.2 Australia's National Training Framework

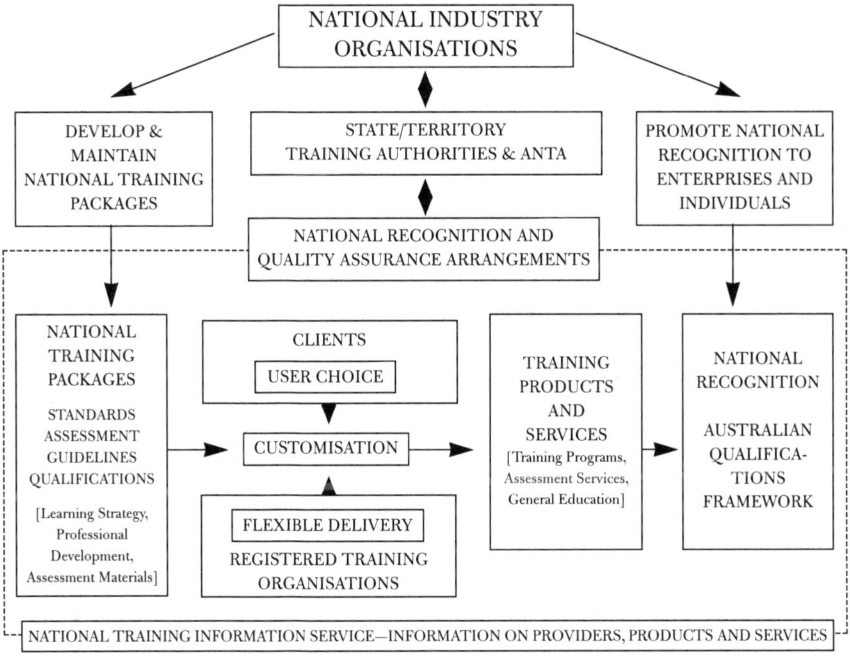

Source: Australian National Training Authority, 1997d

ARF focuses on registration as the basis for improved quality assurance in the VET system. Registration links to national training packages are underpinned by national minimum standards and agreed operational protocols, including audit and monitoring arrangements. This is a move away from the pre-January 1995 system requiring accreditation of courses. The development of industry-led training packages means that registered training organizations (RTO) have the capacity to develop responsive and flexible approaches to delivery that will achieve national outcomes. Other features include effective mutual recognition and improved national consistency (see Figure 3.3, *right*).

The Australian Recognition Framework

As of January 1, 1998, ARF significantly shapes the broad national environment in which skill standards and qualifications (training packages) operate. A comprehensive approach to national recognition of VET, ARF is based on the quality-assured approach to the registration of training organizations that deliver training, assess outcomes, and issue qualifications. ARF seeks to establish a streamlined, responsive system of national recognition underpinned by strengthened quality assurance. ARF's key objectives are to

- support mutual-recognition arrangements between states and territories, RTOs, and industries;
- enable the implementation of training packages and the introduction of more flexible, client-oriented delivery arrangements;
- increase the capacity of recognized organizations to take responsibility for their operations by moving decision-making processes to where client-supplier interactions take place;
- enable user choice and new apprenticeships in the training market;
- integrate VET recognition systems with wider state and territory quality arrangements;
- strengthen the currency of competencies as the basis for recognition within VET.

Figure 3.3 Three Phases of the Development of Australian Skill Standards and Qualifications

Phase One: Pre-Competency Standards

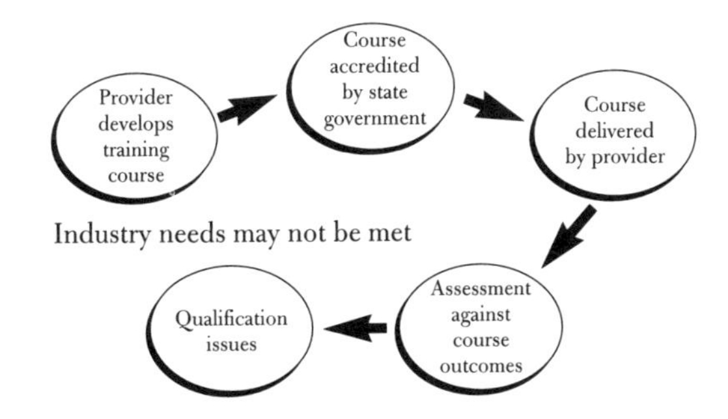

Phase Two: Introduction of Competency Standards

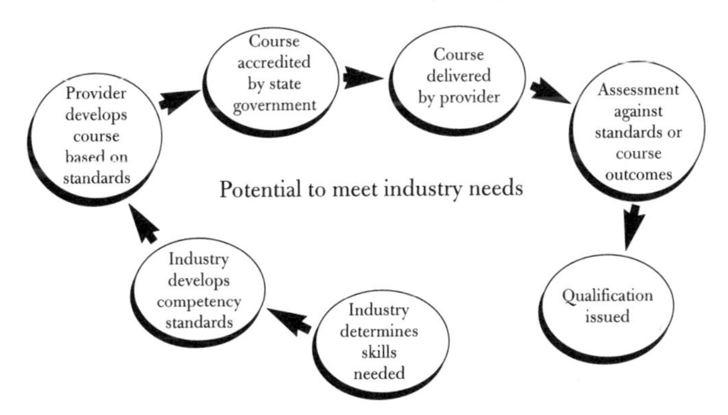

Phase Three: Training Packages

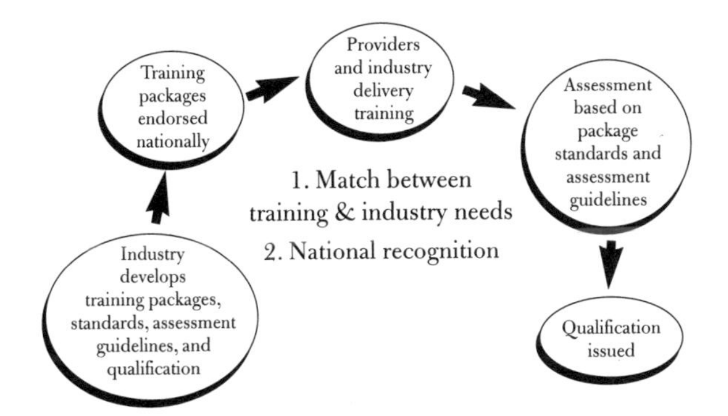

Source: Australian National Training Authority, 1998

The principal mechanism of recognition under ARF is registration of a training organization. The quality assurance cycle for registration has four parts: initial registration, self-assessment and evaluation, compliance audit, and re-registration. These processes will vary depending on the nature of the registration.

National principles, standards, and protocols underpin the operation of mutual-recognition processes by state and territory recognition authorities. The application of mutual recognition is within the context of the particular legislative, occupational licensing, reporting, and accountability requirements of each state and territory. RTOs offering services for overseas clients also need to meet the requirements of the Education Services for Overseas Students (ESOS) Act of 1991.

CURRICULUM STRUCTURE

There has been a reduced emphasis on national or state-level curriculum to increase the scope for flexibility and innovation. While learning resources are still available, their use is optional. Training packages now specify objectives (industry standards) and outcomes (qualifications), but allow flexibility in the delivery of these objectives and outcomes. This is to encourage varied learning pathways. Curriculum content is the responsibility of individual providers, within constraints of nationally agreed outcomes. Unlike the previous framework, assessment is based on industry standards rather than learning outcomes from the curriculum.

National Training Packages

National training packages integrate nationally recognized and endorsed skill standards and qualifications. In Australia, these include competency standards, assessment guidelines and qualifications, and nonendorsable components of learning strategies, assessment instruments, and professional development materials. NTPs deliver current and refined standards, assessment guidelines, and qualifications frameworks—all validated by industry and relevant stakeholders. They provide national training resources that RTOs can use as a basis for developing training programs in specific industry sectors. These nationally available training packages enable the development of tailored training programs to meet particular enterprise or regional needs. This provides options for structured training to reflect student, employer, and training organization needs, while maintaining the core integrity of national qualification outcomes. Training packages also identify apprenticeship and traineeship pathways.

Learning Experience Structure in Training Packages

Training packages enable more variety in pathways and specify qualifications more clearly than in the past. This provides opportunities to customize competency standards to the needs of industries and enterprises, while still maintaining quality of outcomes. All qualifications developed have relevance to the workplace. They can be achieved through contracted employment-based training, unless industry specifies that this is not an option.

Training packages consist of endorsed components (competency standards, aligned to the qualifications framework and assessment guidelines), and nonendorsed components. This structure provides a framework for analyzing the educational experiences provided through the training packages and their contribution to career pathways.

Industry and enterprise involvement in the skill standards approach allows for recognizing outcomes at the unit level through a statement of attainment. This ensures that skills are portable between enterprises and across geographic regions. Clear specification of the elements, performance criteria, range of variables, and evidence requirements for each unit of competence helps achieve this. Each competency standard has set elements and performance criteria, but the range of variables and evidence guides can be customized. This is to ensure the educational experiences are relevant to the trainee and the workplace. Evidence guides specify comprehensive underpinning knowledge and skills to ensure full transferability of learning.

Standards also covering key competencies and language and literacy requirements reinforce this.

The assessment guidelines developed by industry allow for acquiring skills through a range of pathways. Tailoring pathways to the requirements of enterprises and individuals involves partnerships between RTOs and enterprises. In addition, there can be skills learning on or off the job, with a training organization or through a workplace assessor, both of which can provide assessment. A workplace assessor may be registered through an industry-based assessment scheme.

Through the development of training packages, industry has identified multiple levels of qualifications that provide for career progression from entry-level certificate to diploma and advanced diploma. These can be achieved through traineeships and apprenticeships in the workplace, in an off-the-job institution, or in some other combination. This packaging ensures that the requirements for a wide range of pathways are available to achieve a qualification at each level.

The work undertaken on nonendorsable components identifies and develops necessary learning strategies and materials. Doing so identifies, assesses, and uses existing resources. The learning process to achieve the competency standards is more flexible than in the past to encourage innovation and customization for the needs of students and employers. Model training programs, describing a range of learning and assessment pathways, provide a structure to the learning process. Trainer and trainee manuals support training delivery; workplace trainer, coach, or mentor guides are also often available. Together these sets involve the players in a coherent educational framework that allows the trainer, trainee, and workplace supervisor or mentor to achieve optimum conditions and learning outcomes.

New Apprenticeships

A new apprenticeship is an apprenticeship or traineeship involving a formal contract or agreement of training. It offers training flexibilities, support service arrangements, and opportunities for structured training. They can cover full- or part-time work and the part-time arrangements are also available to school students. Businesses can employ apprentices and trainees under Australian Workplace Agreements (AWA), Certified Agreements (CA), or traditional industry awards.

Employment-based training that involves developing skills in the work environment has unique benefits that flow to individuals, firms, and the national skills pool. At the national level, there are eight key issues that are essential to the successful implementation of new apprenticeships. Agreement on these issues early in 1997 enabled states and territories to put their implementation systems in place.

New apprenticeships involve paid work and structured training. Underpinned by a training agreement registered with the relevant state or territory training authority, they lead to nationally recognized qualifications. The training agreement contains a training program that establishes the link between employment and training. There should be an integration of work and training, with time in the workplace being a vital part of training. RTOs play a key role in helping employers choose a training program that suits both businesses and employees.

Under the apprenticeship system of the past, 13 weeks was spent in off-the-job training over a 12-month period. This arrangement traditionally consisted of two days per week, or, in some circumstances, a block release format of several weeks at a time. Apprentices generally spent the same time in training each year during their apprenticeship. As training packages become available, along with User Choice, there will be scope for greater flexibility in delivery arrangements, both on and off the job, to suit the needs of particular enterprises. More flexibility is being added to the traineeship system, including more options for fully on-the-job traineeships.

The Relationship between Training Packages and New Apprenticeships

Training packages have a key role to play since all states and territories (excluding New South Wales) are ceasing the declaration of vocations, which was previously a way for recognizing government-supported apprenticeships. The endorsement of a training package, registration of a training provider, and signing of a training agreement will be the preconditions for funding new apprenticeships.

This expansion of structured entry-level training opportunities will involve small businesses, such as licensed hotels, restaurants, and fast food outlets. Monitoring the effectiveness of the new pathways for small business is a high priority. With new apprenticeships, employers can recruit an apprentice or trainee, select a training provider, or develop a training program. Apprenticeships operate under a formal training agreement between an employer and an employee and are registered with the state or territory training authority. Employers need a registered training agreement to attract public funding and financial incentives and to ensure legal protection.

Trainers, like other businesses, face the discipline of competition. During the employee training time, the training provider is considered an important business partner. Employers assess providers' capacity to meet their needs and the needs of the apprentice or trainee, and to work closely with business. Some providers will be more suited and responsive to employer needs than others, as is the case with any type of supplier.

Aeroskills: A Training Package Example

The Aeroskills training package covers competencies used by people employed in the maintenance and repair of aircraft and components. The training package applies across the industry, including the major national and international carriers, regional carriers, defense forces, and general aviation. The manufacturing, engineering and related services (MERS) industry training advisory body (ITAB) has concentrated on developing and refining a training package for a competent base-level aerospace technician. This is at the level expected at completion of an aerospace apprenticeship in avionics, mechanical, and structures streams. Nonendorsed components of this training package include a learning strategy, assessment materials, and professional development materials for standards initially included in the training package. These components of the training package use existing and newly developed products.

The Aeroskills training package provides outcomes at Certificate II in aircraft avionics, aircraft mechanical, and aircraft structures, achieved through traineeships. Also, there are technician-level (referred to in Australia as trade-level) outcomes achieved through apprenticeships. In this sector an apprenticeship was formerly the primary pathway of learning. Certificate II outcomes provide full articulation with, and credit transfer into, relevant trade qualifications at levels III, IV, and diploma and advanced diploma levels. The Certificate II content is simply a subset of the outcomes of the trade level, providing a qualification outcome for partial completion of the trade training. The training package is designed to accommodate the existing trade apprenticeships of Aircraft Maintenance Engineer (Avionics), Aircraft Mechanic (Avionics), Aircraft Maintenance Engineer (Mechanical), Aircraft Mechanic (Mechanical), and Aircraft Maintenance Engineer (Structures Maintenance).

The training package approach broadens access to public funding, especially for industries looking toward contracted training. Training packages design qualifications for real work outcomes. Development of sub-trade qualifications is encouraged to build skills more widely in the workforce and provide clearer steps to a trade outcome. Industry validates these outcomes to ensure they relate to current needs. Sub-trade qualifications also provide a recognized outcome for people who do not complete the full training for a trade but do have worthwhile and portable skills.

ROLE OF PUBLIC AND PRIVATE EMPLOYER AND EMPLOYEE GROUPS

ANTA draws on the input and needs of the stakeholders in the national VET system and provides the impetus and framework for national strategic direction and coordination. The employer role has increased over the past five years. ANTA's current structure provides for a role for employer and employee groups through the composition of the ANTA board, which consists of people from industry and employer groups. The current board includes representatives from the peak union body, the Australian Council of Trade Unions, and executives from peak industry bodies. These groups include the Metal Trades Industry Association and the Australian Chamber of Commerce and Industry, along with key industry executives.

States and Territories

States and territories within Australia produce yearly state training profiles. These establish a single comprehensive plan for the provision and support of VET in the states and territories. This is based on directions in the national strategy and priorities set by the ANTA Ministerial Council. These profiles are planning and resource documents, forming the basis for the allocation of commonwealth funding under the ANTA agreement. They provide an outline of training activities for the application of commonwealth, state, and territory funds.

Industry Advice

Industry leadership and ownership is critical to build a stronger VET system. In 1996, ANTA reviewed and consolidated the roles and functions of the two main support mechanisms for industry involvement in VET. ANTA supports a national network of industry training advisory bodies. It provides funds to the states and territories to complement resources supporting their industry advisory arrangements. Following a review of industry advisory arrangements, the roles and function of national ITABs were confirmed to include the following:

- developing and maintaining training packages;
- marketing the benefits to industry and enterprises of recognized national training and qualifications;
- providing ANTA and government with strategic advice on industry's VET priorities.

ITAB boards and training package development reference groups usually also have union representation.

User Choice

In 1996, consultation and pilot testing led to the refinement of the User Choice policy. User Choice increases the responsiveness of the VET system to the needs of clients by encouraging a direct and market relationship between individual providers and clients. In July 1996, ministers "agreed to progressive implementation of User Choice during 1997, and to full implementation of User Choice for off-the-job training for apprentices and trainees from January 1, 1998." (This agreement was later clarified to apply only to User Choice implementation for new trainees and apprentices registering after January 1, 1998. New South Wales has reserved its position on the implementation of User Choice.)

Under User Choice arrangements, employers and their apprentices or trainees have the right to exercise choice over which registered training organization delivers their off-the-job training. Employers and apprentices or trainees can also negotiate with RTOs on specific aspects of training within the requirements of the training package, such as training content, timing, and mode of delivery.

In September 1997, ministers agreed that states and territories would implement User Choice for apprenticeships and traineeships in line with ANTA's *Report to MINCO on the Implementation of "User Choice."* The policy statement outlines the objective of User Choice, defines its essential elements and principles, and explains how User Choice will operate beginning January 1, 1998. The policy's principles are as follows:

1. Clients are able to negotiate their publicly funded training needs.
2. Clients have the right of choice of registered provider, and negotiations will cover choice over specific aspects of training.
3. User Choice operates in a national training market not limited by state and territory boundaries.

4. The provision of accurate and timely information about training options is necessary.
5. Pricing of training programs by state and territory training authorities should reflect clearly identified program costs and have reference to unit cost benchmarks.
6. The client can negotiate and purchase customization above what is funded publicly.
7. Regulatory frameworks and administrative arrangements relating to VET at the national, state and territory level are complementary to the achievement of User Choice objectives.
8. Evaluation of outcomes of User Choice against objectives is an integral element of a program of continuous improvement. This requires innovation to achieve and maintain a best-practice training system.

Industry Relationships

ANTA helps industry organizations gather and maintain strategic advisory services and develop and achieve endorsement for national training packages. Other activities ANTA supports include special initiatives in areas such as staff development, front-line management, small business, and the National Transition Program, which involves the conversion of apprenticeships and traineeships into a competency-based format.

ASSESSMENT APPROACHES AND STUDENT ACHIEVEMENT

In recent years the focus of training has moved significantly to outputs or outcomes—the skills or competencies obtained from training. These outcomes are the skill needs of enterprises, defined by industry or enterprises and expressed as competency standards. The focus on the learning process is for flexibility and various pathways of learning and assessment. The system is changing; in the future it will give more authority to providers to respond to clients with delivery and assessment strategies suited to the enterprises and trainees.

The National Training Framework Committee has responsibility for the endorsement of training packages developed by industry groupings or enterprises. These training packages include assessment guidelines specifying the approaches industry believes are best suited to their training needs and delivery strategies. The assessment guidelines may indicate preferred levels of assessor qualifications that state training authorities may take into account as part of the training organization's registration process. Within this framework assessment can occur on or off the job by an organization registered for training and assessment services or for assessment services only. Employers and trainees, therefore, have a maximum opportunity for skills recognition and the attainment of qualifications through a range of training and assessment pathways. Many industries are considering developing a "skills passport" to record trainee achievement for the benefit of both trainees and employers in a dynamic and mobile workforce. The focus on industry-based competency assessment has also meant that professional bodies, such as the Institution of Engineers, have the opportunity to recognize skills at subprofessional levels.

OPPORTUNITIES FOR EMPLOYEE CONTINUING EDUCATION

Middle-Level Skills

Raising the Standard, a 1993 report on middle-level skills in the Australian workforce (that is, skills possessed by people in technical, administrative, supervisory, or newly emerging occupations), recommended expanding these skills to meet the demands of industry and the global economy. Strategies to attain this goal included developing broader recognition of prior learning programs and lifelong learning strategies that would raise workforce standards and meet demand for these skills (National Board of Employment, Education, and Training, Employment and Skills Formation Council 1993). The report also focused on looking at a variety of ways to attain qualifications, rather than the traditional TAFE curriculum model. Today, there is a return to industry and workplace assessment and training, one of many delivery pathways to qualification outcomes.

Adult and Community Education

The ACE sector is one pathway to assessment that provides lifelong learning opportunities for adult Australians. Each year, about one million people take ACE courses. The ACE Taskforce of the Ministerial Council on Employment, Education, Training and Youth Affairs advises ANTA about priorities for the sector. The taskforce has representatives from ANTA; commonwealth, state, and territory governments; and the Australian Association of Adult Community Education.

A consultancy project analyzing the role of ACE in VET ran in conjunction with the task force and managed by ANTA. The ACE sector widely considered the consultancy project report *Think Local and Compete* (Australian National Training Authority 1996), a valuable analysis that helped the taskforce revise national ACE policy. The coordination of Adult Learners Week nationally and support for the Australian Association of Adult and Community Education further promoted the opportunities for development and recognition of middle-level skills.

APPROACHES FOR TEACHER PREPARATION

Australia has more than 2,500 registered providers and there are strategies to enhance teacher preparation. In TAFE colleges, teachers are required to have industry experience and complete TAFE teaching qualifications. Staff in any RTO involved in assessment toward formal qualifications must be competent in the technical skills at the level being assessed and also in national Workplace Assessor competencies.

The new National Training Award for Training Provider of the Year is offering greater recognition of best practice. The introduction of the Demonstrating Best Practice in VET project also encourages teachers to pursue continuous improvement. There are pilots of User Choice for apprentices and trainees, with $108 million of public funding in 1996 used to purchase places on a contestable basis. Reduced regulation and a focus on guaranteeing outputs has also been a key part of the strategy to improve teacher preparation.

FRAMING THE FUTURE

Framing the Future is a major national staff development initiative funded through ANTA that uses work-based learning to promote understanding of and participation in the National Training Framework. This project aims to provide a group of skilled, well-informed advisers who can support NTF implementation in their organizations. It offers a range of products to promote NTF and is supported by an Internet site. Designed for people in the VET sector, Framing the Future features a series of work-based learning projects in all states and territories, plus the formation and maintenance of a national advisor network.

Framing the Future provides limited funding to support work-based learning projects. Experience has shown that staff development using work-based learning principles is the most effective way of providing relevant, timely, and efficient skills acquisition for participants. These projects provide opportunities for organizations to address some specific NTF implementation issues. During 1997, there were 40 work-based learning projects conducted throughout Australia, with over 600 participants. In 1998 there are 50 projects with a focus on ARF and implementing training packages.

SYSTEM EFFECTIVENESS

In 1996, the Australian VET sector achieved the following:

- More than 1.35 million students (of whom 47 percent were female) were in VET programs delivered through public funds combined with fee-for-service provision through public providers.

- Public funds delivered more than 263 million annual hours combined with fee-for-service provision through public providers, an increase of 2.5 percent from 1995.
- There were 161,496 apprentices and trainees under contracts of training as of December 31, 1996, an increase of 17 percent from December 31, 1995.
- The participation of Aboriginal and Torres Strait Islander peoples in VET increased from 2.1 percent in 1995 to around 2.4 percent in 1996. Indigenous peoples represent 2 percent of the total population.
- Seventy-four percent of employers responding to the National Employer Satisfaction Survey indicated satisfaction with VET programs.
- Large employers expressed satisfaction with the skills acquired by and the job readiness of graduates (86 percent and 84 percent, respectively).
- Around 80 percent of TAFE graduates who wanted to get a job were satisfied with their course outcome.
- Almost 85 percent of students in VET modules (subjects) achieved a successful competency or were in the process of attaining competency.
- The overall student pass rate was 82.2 percent and the overall completion rate was 85.1 percent.
- Government at commonwealth, state, and territory levels spent approximately $2.79 billion on VET.

KEY ISSUES AND TRENDS BEING ADDRESSED FOR IMPROVEMENT AND FOR ADJUSTMENTS

Broadly speaking, the new National Strategy for 1998–2003—A Bridge to the Future—outlines key issues facing VET. The strategy focuses on the following areas:

- equipping Australians for the world of work;
- enhancing mobility in the labor market;

- achieving equitable outcomes;
- increasing industry investment in training;
- maximizing the value of public expenditure on vocational education and training.

The direction in setting these objectives is to create an environment where postsecondary education and training opportunities are almost universal in Australia. Having these objectives means that VET can be delivered in ways that meet differing individual needs, rather than simply through traditional courses. The objectives could also lead to skill standards recognized and valued by employers anywhere in Australia. The underlying purpose of these goals is to help the Australian workforce gain international recognition for the quality of its skills.

Qualifications. Equipping Australians for the world of work is a key emerging issue for VET in Australia. Qualifications—in full or part—are the principal means of formally recognizing competence. In many industries, however, qualifications are not available and many existing qualifications do not adequately signal to employers the skills acquired by individuals. People have often gained competency through informal learning that has not been officially recognized.

There will be a wholesale review, rationalization, and upgrading of existing standards and qualifications and the development and extension of standards and qualifications into many new industry sectors. This will provide opportunities and experience for a much broader range of people than in the past with outcomes easily recognized and valued by employers anywhere in Australia.

Skills recognition. Many people enroll in training programs to achieve competence in selected areas, rather than full qualifications. Through workforce skills recognition, it is possible to achieve assessment of many people as competent in specific areas. Recently, agreement has been reached for a consistent approach to issuing "statements of attainment." Presently, the vocational education and training system cannot recognize these outcomes in a nationally consistent manner, a situation that restricts the mobility of the workforce. As these new qualifications become available, new and different learning pathways must be available for people wishing to access them.

Traditionally, the main way to get a VET qualification was by either a TAFE or private provider course or an apprenticeship (and more recently a traineeship). As valued as these pathways were, they did not suit everyone. Apprenticeships were largely restricted to young men and only available in a narrow range of industries and occupations. Also, industry and individuals are looking for more choices about what, where, when, and how they learn. The VET sector in Australia will continue to build on policy reforms of the recent past to offer individuals a far greater choice of pathways to a qualification. As they pursue a qualification, individuals might do structured training at an institution, on the job, through new apprenticeships, at secondary schools, in community settings, or in the home, or through a combination of two or more of those methods. Alternatively, they might not do any structured training at all; rather, they might have their current competencies assessed against the requirements of a particular qualification.

Unrecognized training. An enormous, unquantified amount of unstructured, unrecognized training occurs in every workplace, every day. Many people rely on this sort of informal learning at work to build their competence. Sometimes, they also do formal training on the job. Reforms to the assessment of skills will enable people who receive both formal and informal on-the-job training to have their current competence recognized toward a national qualification. There will also be an increase in the amount of formal training provided on the job.

School role. Many students, parents, and schools want opportunities for students to get work skills and experience during secondary school, and to receive credit for it. They want clear pathways into jobs. Previously, secondary schools regarded their main role as preparing young people for access to higher education. In the future, the quality and quantity of nationally recognized vocational education and training in schools will increase.

Online technology. The use of online technology can provide or supplement all the new learning pathways. There are 1.4 million registered Internet users in Australia. From 40 to 50 percent of Australian homes and most workplaces have personal computers. Therefore, the option of delivering training using the Internet and new technology is available to more people than before. To achieve this, however, there will have to be more material available for online use.

There are currently more than 10,000 nationally recognized courses and more than 2,500 RTOs. Clients do not have easy access to information on courses and RTOs. Conversely, RTOs find it difficult to access both standards and nationally recognized courses. The National Training Information Service (NTIS) is helping to provide information on the Internet about industry competency standards, registered training organizations, training packages, and available courses.

Literacy. Individuals who have poor literacy and numeracy skills find it harder to get a job than do those with better skills. It is also harder for them to get training and qualifications to make them mobile in the workforce, even if they are competent at their work. Some 6.2 million Australians between the ages of 15 and 74 have poor or very poor literacy and numeracy skills. Training programs will, if necessary, teach people language, literacy, and numeracy skills at the same time as teaching them industry competence. In addition, there will also be stand-alone courses for those who need them.

Australian industry currently spends around $3 billion on all forms of education and training, although a great deal of this spending is not formally recognized. Despite this, research indicates that training provided by employers is not increasing at the rate estimated to be necessary to maintain the competitiveness of Australian industry. Employers treat expenditure on training as they do any other investment decision—the level is directly related to the perceived return.

Overall economic conditions also influence investment in training. The growth in the availability of products developed and valued by industry and the various services available in forms that directly meet the needs of individual enterprises should lead to an increased investment by industry in recognized training. Governments can directly influence industry investment by making available resources to leverage industry expenditure. They can also encourage partnerships between RTOs and enterprises in areas such as skills recognition and assessment.

Industry investment. Industry investment in training should not be seen purely in terms of expenditure. Work management practices also heavily influence learning and skills acquisition in the workplace. These include initial recruitment, the identification of competencies as the clear basis for promotion and career paths, skills audits, job design, and work organization to promote multiskilling, skilling, and employee counseling and guidance. As an integrated set of benchmarks and resources, training packages can play a useful role in assisting firms with these processes.

Underrepresented groups. VET participation and completion rates for women, Aboriginal and Torres Strait Islander peoples, and persons with disabilities have been significantly different from those for the population as a whole. While their participation has been around the average, women concentrate in a narrow range of fields. It is typical to cluster indigenous Australians at the lower end of the qualification's spectrum and successful completion of programs remains a problem. People with disabilities also tend to do preparatory courses, or courses with low or declining labor market demand.

Many of the reforms and initiatives that are already under way in the VET sector have the potential to open up opportunities and pathways for people who are not well served by current arrangements. It is necessary for a full review nationally and in the states and territories to identify legislation, regulations, and practices that restrict the capacity of RTOs and other service delivery agencies to respond to the needs of a diverse range of clients—especially those underrepresented in the system.

Government investment. Government expenditure on vocational education and training now totals $2.5 billion. Governments and taxpayers expect maximum value from that expediture. The ANTA Ministerial Council has recently made two important decisions in that regard. It endorsed a framework by which more training will be provided by the system. Also, states and territories will identify strategies and outcomes (both quality and quantity) to achieve growth through efficiencies in their systems. These strategies will be specific to their individual circumstances and histories of efficiency improvement. There are other measures under way that will also improve system efficiency.

Training facilities. More than $4 billion has been spent on providing a comprehensive network of public vocational education and training facilities in the last 25 years. There must be a maximization of the use of these facilities. In some states and territories, access to public training facilities by other providers is now possible. States and territories are shifting their capital investment from "bricks and mortar" to infrastructure that supports flexible delivery—particularly in the use of new technology. However, a full examination of the use of current capital stock and the basis of future capital investment is required.

Information collection. In recent years, steps have been taken to improve the quality of the information available to vocational education and training managers and clients, and there are plans to collect better information, including through surveys of graduate destinations and employer and student satisfaction. This information will help providers improve their ability to help clients make better decisions. ANTA will report some of this information in its annual national report. Governments will expect public RTOs to improve their commercial performance through techniques including accrual accounting, output costing, activity-based financial information, local and international benchmarking, and the establishment of competitive neutral cost bases.

CONCLUSION

The Australian system is shifting its focus toward a more flexible industry-based competency assessment. The system is seeing an expansion in the number of options for the provision and delivery of training, including online and remote-access training, along with the more traditional sources of training such as TAFE. The key issues here are multiple learning pathways and flexible delivery. There is less emphasis placed on the curriculum and more focus on training outcomes.

As in the United States, the Australian VET system is becoming more standards based, with a comprehensive set of standards set by industry. Commonwealth, state, and territory governments and ANTA facilitate these standards, and assessment and qualifications are subsequently based on them. This ensures that industry gets the standards and qualifications it considers necessary for improved productivity. The setting of these standards will also see a gradual shift to a national VET system, removing the artificial barriers of states and territories. The Australian system addresses federal structures and stakeholders in VET to create a flexible system that is focused on achieving industry outcomes that will improve Australia's productivity.

ABOUT THE AUTHOR

Peter Noonan is the acting chief executive officer of the Australian National Training Authority. Before his work with ANTA, he was involved for more than 12 years in vocational education and training nationally and in Victoria. Late in 1993, Mr. Noonan received a Fulbright fellowship that allowed him to study

U.S. reforms in vocational education and training at the National Center on Education and the Economy in Washington, D.C.

BIBLIOGRAPHY FOR CHAPTER 3

Australian Bureau of Statistics. 1996. *Aspects of Literacy: Assessed Skills Levels.* Belconnen, Australian Capital Territory: Australian Bureau of Statistics.

Australian Committee on Technical and Further Education. 1974. *TAFE in Australia: Report on Technical and Further Education, Volume 1.* Brisbane, Queensland: Australian Committee on Technical and Further Education.

Australian National Training Authority. 1994. *Towards A Skilled Australia. A National Strategy for Vocational Education and Training.* Brisbane, Queensland: Australian National Training Authority.

———. 1996. *Think Local and Compete.* Brisbane, Queensland: Australian National Training Authority.

———. 1997a. *Annual Performance Report. 1996–1997.* Brisbane, Queensland: Australian National Training Authority.

———. 1997b. *Benchmarking Vocational Education & Training.* Vol. 3, Annual National Report. Brisbane, Queensland: Australian National Training Authority.

———. 1997c. *Commonwealth, State & Territory Achievements.* Vol. 2, Annual National Report. Brisbane, Queensland: Australian National Training Authority.

———. 1997d. *National Overview.* Vol. 1, Annual National Report. Brisbane, Queensland: Australian National Training Authority.

———. 1997e. *Report to MINCO on the Implementation of "User Choice."* Brisbane, Queensland: Australian National Training Authority.

———. 1998. *A Bridge to the Future—Australia's National Strategy for Vocational Education and Training 1998–2003.* Brisbane, Queensland: Australian National Training Authority.

———. 1998 "ANTA's Charter." Internet: www.anta.gov.au/about/default.htm.

Cullen, R. B. n.d. "Workskills and National Competitiveness Report No. 3; Internal Benchmarks." n.p.

Department of Employment, Education, Training, and Youth Affairs. 1996. *Selected Higher Education Statistics.* Canberra, Australian Capital Territory: Department of Employment, Education, Training, and Youth Affairs.

National Board of Employment, Education and Training, Employment and Skills Formation Council. 1993. *Raising the Standard: Middle Level Skills in the Australian Workforce.* Canberra, Australian Capital Territory: Australian Government Publishing Service.

National Center for Vocational Education Resources. 1997. *Australian VET Statistics.* Adelaide, South Australia: National Center for Vocational Education Research.

Chapter 4

The Vocational and Technical Education System in Japan: Technicians for Today and Tomorrow

Elizabeth J. Teles
National Science Foundation, Washington, D.C.
Hidetoshi Miyakawa
Aichi University of Education, Aichi, Japan

Since the mid-1950s, there has been a strong demand for technicians, technologists, and engineers in Japan who can cope with the advancing industrial technology resulting from Japan's progress in science and technology. To help meet this demand, the Ministry of Education, Science, Sports, and Culture (Monbusho) has established a college of technology in each prefecture (state) to offer associate degrees in engineering and science technology. According to Monbusho, in the colleges of technology:

> The emphasis in the educational program is on improving basic knowledge, thus providing students with the ability to respond efficiently and creatively to the increasing complexity of science and technology in the modern world. In addition, a high educational standard is maintained through ample opportunities to carry out experiments and practical training in their chosen field of specialty (Association of National Colleges of Technology 1993).

This mixture of providing basic education combined with significant training in specific areas characterizes all technician education in Japan at many types of institutions.

THE SCHOOL SYSTEM AND VOCATIONAL AND TECHNICAL EDUCATION

In 1947, Japan enacted the current Fundamental Law of Education and established a 6-3-3-4 system to realize the principal of equal opportunity in education. At present all children between the ages of 6 and 15 attend six years of elementary school followed by three years of lower-secondary education. This system of nine years of compulsory general education is one of the strongest in the world. More than 96 percent of students subsequently attend an upper-secondary school, with about 74 percent of those choosing a general education course of study and about 26 percent choosing a specialized course focusing on vocational education in such areas as agriculture, industry, business, fishery, home economics, or nursing. The percentage of upper-secondary students selecting vocational education is down from about 40 percent in 1970. Part of educational reform in Japan today is attempting to reestablish an attractive, interesting, and valuable vocational upper-secondary school system to meet the diverse needs of Japanese society. For example, in 1994 a few integrated secondary schools were created to provide vocational education students with both general and specialized courses of study as elective subjects. The integrated system has a much more flexible curriculum and uses a credit system.

Monbusho creates national curriculum standards for all school levels, from kindergarten to upper-secondary schools. This powerful ministry determines most of the educational policy for all types of educational institutions in Japan. There is truly a national system of education. Guidelines for objectives and standards are prepared by Monbusho, reviewed by the Curriculum Council (an advisory body to Monbusho), and announced by the minister of education, science, sports, and culture. While there are some local differences, Monbusho ensures that the system of education is national and courses of study vary little among schools. In vocational schools in particular, the fact that the laboratory equipment, computers, and machinery that students use are the same in each of the secondary schools offering similar programs serves to ensure homogeneity among programs.

About 46 percent of upper-secondary school graduates attend one of the universities, junior colleges, or colleges of technology that form the basis for the system of higher education in Japan. All national universities, local public universities, private universities, junior colleges, and colleges of technology are under the administration of Monbusho, although each institution has its own governing body. In 1974, Monbusho authorized under a special law the creation of special training colleges that afford growing access to specialized courses of study for many young Japanese students, increasing to almost 65 percent the number of upper-secondary school graduates who receive some type of higher education. In addition to public and private colleges and universities, companies in Japan offer specialized upper-secondary and college education that leads to one-year certificates or two-year associate degrees, bringing to over 70 percent those receiving additional education or training after upper-secondary schools. Company schools and colleges are authorized by and come under the administration of either Monbusho or the Ministry of Labor.

NATIONAL ADMINISTRATIVE STRUCTURE: TYPES OF ENGINEERING AND SCIENCE TECHNICIAN EDUCATION

The description above sets the context of vocational and technical education within the broader system of education in Japan. The remainder of this essay concentrates on the education of science and engineering technicians. Most of technician education still occurs at the secondary school level; however, there is a growing trend toward options in higher education for students selecting a vocational secondary school track. In addition, about 20 percent of Japanese undergraduates are majoring in engineering or engineering technology and about 4 percent are studying science. Many of these engineering students are at the national institutes of technology, where an engineering major is roughly equivalent to an engineering technology major in the United States.

There are many varied institutions where the education and training of science and engineering technicians for the high-performance workplace occur. The data that follow come from Monbusho (Ministry of Education, Science, Sports, and Culture 1997b). Among the institutions offering science and engineering education and training are the following:

Industrial secondary schools. These account for 837 secondary schools with about 402,620 students out of 4,539,694 secondary school students, or about 9 percent of the upper-secondary school population (Monbusho schools).

Integrated secondary schools. These schools serve 14,147 students, or about 0.3 percent of the total secondary school population (Monbusho schools).

Industry-associated schools. These secondary schools are associated with and run by industries such as Toyota and Denso. For example, the 439 technical high school students at Toyota are employees who earn salaries while they attend school.

Colleges of technology. These 62 colleges of technology have five-year programs and about 56,396 students, of which 34,147 are at the level of upper-secondary school and 21,608 are fourth and fifth year students at the associate level (Monbusho colleges, with one in each prefecture plus a few additional specialized ones).

Specialized institutions. These specialized institutions are run by industries but grant certificate and associate degrees.

Junior colleges. 463,948 students attend junior colleges that grant associate degrees; 21,357 students, or 4.6 percent, are in engineering fields (Monbusho schools).

Technical institutes. The technical institutes grant associate degrees and are organized by authority of the Ministry of Labor; they include about 20 institutions, each of which has about 400 students, for a total of about 8,000 students. Students graduating from the institutes have little chance to transfer to another type of institution (Ministry of Labor colleges).

Specialized training colleges. These training colleges are authorized by Monbusho under the Special Education Act.

Private training institutions. Various other institutions are not under the authorization of Monbusho.

CURRICULUM STRUCTURE

Specialization of Institutions and Tracking of Students

In contrast to the United States—where most technician education takes place in two-year colleges, through tech prep consortia between high schools and two-year colleges, or through the military—most of the technical upper-secondary schools and colleges in Japan offer only a few programs. They range from one in automotive technology at Toyota Associated High School to nine at Tokyo Polytechnic Institute. Japanese programs are highly specialized and resemble more the private types of associate degree institutions in the United States, such as Devry, than the two-year technical associate degrees in U.S. public colleges and high schools. Most of these high schools and colleges have fewer than 500 students.

Most students preparing to be technicians make career choices in the ninth grade. Few changes are made after age 15, since there is little movement among programs. Students who graduate from secondary school in Japan go on to a university or college (including colleges of technology, technical institutes, private specialized colleges, company-operated colleges, and junior colleges) or immediately to employment. Since as in upper-secondary schools there is little movement between jobs or among programs in colleges and universities, virtually all students decide by age 18 what they will be doing for their lives. College students are predominantly aged 18 to 22. Except for some continuing education that is beginning to assume more importance in Japan, most formal education takes place before students ever enter the workplace. Because of the importance of the national examination, the centralized funding of higher education, and the setting of student quotas in departments, there are few transfers among programs or institutions. One result of this approach is almost all students who begin programs finish. On the other hand, it also means that few students ever change careers once they enter employment or a college program.

General and Specialized Courses

At the technician level, training is important. Upper-secondary schools all have general courses that students take. These courses include mathematics, science, social studies, and Japanese (and in some cases English), but all programs encompass many hours of specialized training. As students near graduation, more and more emphasis is placed on specialized courses to prepare students for employment.

At the associate degree level, mathematics is at a higher level than it is in most associate degree programs in the United States. Secondary school mathematics instruction emphasizes general mathematics skills. At the colleges of technology, mathematics and other general requirements are much higher than at the colleges administered by the Ministry of Labor or the specialized colleges. In the colleges of technology, a student finishing with an associate degree has about the same mathematics preparation as a student completing a four-year college engineering technology program in the United States.

While there is a base of general education courses in all the curricula, the prime emphasis is still on training. All types of institutions designing programs for technicians emphasize preparing students for work. This emphasis, however, seems to vary with the type of institution. Students enrolled in colleges of technology have high academic skills and abilities in mathematics and science; many of them now transfer to four-year engineering technology programs. Those at the Ministry of Labor technical institutes are expected to enter the workforce.

For example, in the electronics program at Denso, students receive an associate degree in mechatronics. The curriculum consists of two years of general study, including differential and integral calculus, Laplace transform analysis, physics, English, and physical education, along with specialized technical training the first year. The specialized technical training includes measurement engineering, electric circuits, electronic engineering, electronic circuits, machine manufacturing, industrial materials, electromagnetics, microcomputer control engineering, automotive electric equipment, automotive engineering, strength of materials, and machine elements. In the second year, the training involves instruction in electric/electronic materials, integrated circuits, electric equipment, microcomputer system design, communications engineering, machine dynamics, and production engineering.

The colleges of technology offer a 3 + 2 (with options for another +2) program. For example, an associate degree in engineering technology at Toyota College of Technology would follow three years at the college of upper-secondary school with options for further university study. Requirements for graduation include general education requirements of 84 credits (31, 23, 14, 11, and 5 credits, respectively, in the five years) and 92 credits of technical education (3, 12, 20, 27, and 30, respectively). The 84 credits of general education (79 required for all plus 5 electives) for all technical associate degrees include 10 credits of Japanese, 16 in mathematics, 10 in chemistry and physics, 10 in physical education, 17 in social studies areas, and 16 in English. Technical subjects for mechanical engineering technology include applied mathematics; applied physics; and general engineering subjects such as courses in materials, manufacturing, mechanics, and electronics. Two credits of internships and eight credits of a special project are required.

ARTICULATION TO HIGHER EDUCATION OR EMPLOYMENT

In order to respond to society's needs and to the desire of students for more options, there is a trend toward the provision of more diversified access to university education. While still rare, a number of reforms have been introduced to achieve these aims, although technical education is still primarily designed to prepare students for work. For example, in 1995 more than 10,000 students transferred from colleges to universities and more than 2,000 from colleges of technology to universities (Ministry of Education, Science, Sports, and Culture 1997b). Both figures are double the totals for 1989.

- Of the 131,000 technical industrial secondary school graduates, about 38,000 (29 percent) go on for additional education (12,000 go to a university or college including junior colleges; 21,000 to a special training college; 5,000 to vocational training). Some 87,000 (66.5 percent) enter employment, and about 6,000 (4.5 percent) are unemployed.

- Of the approximately 10,000 engineering graduates of junior colleges, 1,100 (11 percent) advance to higher education at a university, 7,500 (75 percent) find employment, and 1,400 (14 percent) are unemployed.

- Of the approximately 10,000 graduates of the colleges of technology, approximately 2,500 (25 percent) advance to higher education at a university, 7,300 enter employment (73 percent), and 300 (3 percent) are unemployed. Two national universities, Nagaoka and Toyohashi Universities of Technology, were established primarily for graduates of the colleges of technology. They admit college of technology graduates, offering them continuous education to graduate schools. Some of the colleges of technology have recently been permitted by Monbusho to create their own junior- and senior-level programs.

- Of the students who graduate from the vocational training institutes, only about 1 percent per year qualify for higher education. They must begin at the freshman level.

- Of the approximately 1.5 million secondary school graduates in Japan, about 605,000 (40 percent) enter a university or college; 465,000 (31 percent) enter specialized or vocational training, including special industry training programs; and 366,000 (24 percent) find employment.

There is also growing flexibility in education-delivery systems with the introduction of day and evening undergraduate courses, the introduction of a special register system that provides adults with increased opportunities for study on a part-time basis, and opportunities via the University of the Air. In addition, the National Institution of Academic Degrees can now confer degrees on individuals who have reached a standard equivalent to those who have completed degree programs in universities. Under special circumstances universities can now award credit for study outside their jurisdictions. While this is still rare, it is officially recognized as an important innovation, as noted in the 1995 Monbusho document, *Remaking Universities: Continuing Reform of Higher Education* (Ministry of Education, Science, Sports, and Culture 1995).

ROLE OF PUBLIC AND PRIVATE SECTOR EMPLOYER AND EMPLOYEE GROUPS IN THE NATIONAL SCHEME

Business and industry are involved in the education of technicians in at least two ways. A direct way is through the upper-secondary schools, colleges, and special training schools that have been established and are operated by businesses themselves. The second is through influence in Monbusho.

Technical Training in Business and Industry

Among the many businesses and industries that operate their own industrial technology training centers are internationally recognized firms like Denso Manufacturing, Hitachi, Nippon Electric, Mitsubishi, and Toyota. The Industrial Technology Training Center at Denso, for example, is organized into three divisions:

- The Skill Development Division includes a technical high school course for junior high school graduates and technical skills development courses for national and international Vocational Olympics.

- The College Education Division provides associate degrees in electronics and information processing and college guidance for others who wish to continue to a college education.

- The Skills Improvement Division develops skills training and on-the-job education. A Technical Skills Certification Section gives national and in-house certifications and provides basic technical skills training; and Technique Training Sections provide multi-skilled technicians training and specialized skills training in technical areas like robotics and electronics.

The objectives of high school education at Denso are to provide advanced knowledge of the wide spectrum of technologies applicable to technical needs of Denso Manufacturing and to prepare students for the future of enterprise activities. Advantages for students include completion with a high school diploma, admission to Denso Technical College, and training on state-of-the-art equipment. The electronics associate degree program is designed to train and educate technical engineers (technicians) who can cope with both electronic and manufacturing equipment in a new form called mechatronics, while the information processing technical engineer with an associate degree is capable of intelligently and systemically designing products and computer software.

Hitachi Ibaraki Technical College offers a 15-month education for technical high school graduates who are employed by Hitachi and its group companies. This is an intensive program of 2,240 hours plus extra hours of English conversation class. Four departments offer technical associate degrees in electrical, mechanical, material, and management engineering.

Nippon Electric Technical Training College accepts students for its two-year program after graduation from high school. The college offers students an Integrated Technology Study Program to widen the breadth of knowledge in major technical fields, a Principal Technology Study Program to deepen knowledge in specific advanced technologies, and a Relevant Technology Study Program for rapid improvement in knowledge and capability for their current jobs.

Monbusho

Monbusho has four broad areas of responsibility: education, science, culture, and sports. Its decisions and policies dominate higher education. For example, the Higher Education Bureau within Monbusho is responsible for the

> formulation of basic national plans for higher education; the provision of planning, assistance and advice for the promotion of higher education; the approval and establishment of universities, junior colleges and colleges of technology; planning, assistance and advice as to student welfare and financial aid; the approval of higher education corporations; the provision of guidance and advice to higher education corporations and the promotion of, and provision of assistance to, activities of private educational institutions (Ministry of Education, Science, Sports, and Culture 1996a).

Because of a general concern for higher education in Japan, Monbusho has instructed its University Council to study and recommend specific measures for the advancement, individualization, and revitalization of education and research in universities and other institutions of higher education. The reasons for this charge include progress in scientific research and changes in human resource development needs, a rise in students continuing to higher education and diversification of students, and a growing need for lifelong learning and changing social expectations of universities.

In particular, the business and industry sector has been making many recommendations concerning the future of human resource development. Japan currently is characterized by rapid social and economic changes, including internationalization and a shift to an information society, that require the development of workers who combine advanced knowledge and technical skills with a broad perspective and ability to make judgments. Monbusho has been guided in its new higher education policies by at least eight reports from business and industry groups, including two from the Japan Federation of Economic Organizations and two from the Tokyo Chamber of Commerce and Industry. While the University Council makes the recommendations, these reports have heavily influenced their policy decisions.

Business and industry are also somewhat involved in planning all Japanese policies for education for the next century. In 1996, the Central Council for Education of Monbusho released *The Model for Japanese Education in the Perspective of the 21st Century*. (Ministry of Education, Science, Sports, and Culture 1996b). The council included three representatives from industry among its 35 members.

ASSESSMENT APPROACH FOR PROCESS AND STUDENT ACHIEVEMENT

Entry Requirements for Programs Leading to Associate Degrees

Unlike most two-year colleges in the United States, which are open-door institutions, Japanese institutions have high entrance requirements for technical education programs. At the colleges of technology and the polytechnic institutions, entry requirements consist of locally administered examinations plus interviews. Both types of institution still have many more applications than they can accept.

While the acceptance rate among students who qualify is relatively high, the institutions are still highly competitive, and they can choose the most highly qualified students. However, as in the United States, more and more students wish to keep their options open for as long as possible. In addition, students' desire to become technicians is dropping.

The colleges associated with companies such as Toyota and Denso give examinations and interviews at the high school level. The specialized schools do not have rigid requirements and tend to accept virtually all students who apply. These are primarily open-door institutions, but are private and generally quite expensive to attend.

Exit Requirements for Students

Most technical fields have nationally administered examinations that students must pass to be certified in those fields. Since competency is important, students who attend Toyota, Denso, and similar companies must also pass company requirements. Workers have annual evaluations as well as two other evaluations per year that help determine bonuses and advancement.

Program Effectiveness Assessment

Monbusho and the associated Council of Higher Education are responsible for assessing systems and programs. Monbusho collects and publishes data on the numbers of students who apply to programs, pass entrance examinations, complete programs, and secure employment. Also, Monbusho recently has undertaken studies of student and industry wishes concerning classes and curricula. Monbusho requires colleges and universities to implement various forms of self-monitoring and self-evaluation and to publish reports on their results. In addition, it encourages evaluations by academic societies and associations. The general framework is defined by laws and regulations; however, within that framework, college and university administrators set up their own mechanisms.

OPPORTUNITIES FOR EMPLOYEE CONTINUING EDUCATION

Lifelong Learning

Recently, Japan has adopted an education policy aimed at creating in students a zest for learning and giving them the ability to continue to learn. Paralleling this policy has been the adoption of a national plan that emphasizes lifelong learning. The Lifelong Bureau of Monbusho was formed with the objective of expanding various opportunities to learn and promoting appropriate means to evaluate the results of learning in order to create a society whose members continue to learn throughout life. Although the Japanese government is emphasizing the importance of continuing education, little continuing education is now connected to work-related knowledge and skills. But this area is growing. Most of the successes to date have been in the areas of health, sports, and hobbies, with only about 10 percent of continuing education designed to increase work-related knowledge and skills. Most of what people report is learning through books, magazines, television, private lessons (Juku), newspapers, radio, and tapes. Of the lifelong learning activities reported, 1.3 percent take place at special training colleges, 0.7 percent at universities and junior colleges, 0.5 percent through extension courses, 0.2 percent at vocational training schools, and 0.1 percent at the University of the Air.

Ongoing Education in Business and Industry

A much more common method of stimulating lifelong learning is the encouragement of on-the-job training and education provided by Japanese companies and businesses. Reports show that in 1994 (Ministry of Education, Science, Sports, and Culture 1997b), 51 percent of Japanese workers underwent on-the-job training and 57 percent involved themselves in self-improvement activities. For example, a key component of Toyota's human resource development system is participation in company training programs that supplement on-the-job training. In special skills and knowledge training, employees learn the practical skills and knowledge necessary for performing daily operations. In specialized technical skills certification, which aims at advancing production team members' skills and knowledge, technicians move through four grades of expertise with corresponding responsibilities and salaries.

In 1994, more than 80 percent of Japanese companies and businesses provided some type of support for workers' self-improvement. Support included financial assistance, adjustment of working hours for training and education, sponsorship of courses, and paid educational leave. Worthy of note, however, is that of the 19 percent of workers who took paid educational leave, 62 percent were away for less than three days.

STUDENT RECRUITMENT, RETENTION, AND PLACEMENT

As more Japanese students desire to attend general education high schools to keep their career options open longer, recruitment for the vocational and technical secondary schools becomes more difficult. Many technical secondary schools complain that it is becoming harder to get good students. Toyota, for example, has instituted a broader national recruitment plan to attract students to its technical high school. Monbusho has instituted integrated high schools to give students more course options.

However, most students who enter technical programs finish. This seems to be the result of a combination of several factors, including

- department and college funding depends on the number of students in programs—so it is important that students finish to allow colleges and universities to meet enrollment numbers;
- company hiring policies in general depend on a student finishing a program of education;
- career tracks for technicians let those who get an associate degree get a higher job classification, often as master technicians;

- national certification examinations for most technical jobs prompt schools and colleges to prepare students to take examinations that determine who can practice in many fields;
- the Japanese have a general attitude that people should finish programs that they start, and they take pride in those who do.

As in the United States, highly specialized technical skills and abilities at the technician level are valued by industry. Colleges and secondary schools place virtually all of their graduates in companies, with many more jobs available at this level than there are applicants. While placement does not seem to be an issue, having enough qualified technicians graduating from programs is an issue. In the United States, this shortage can be attributed to a combination of recruiting insufficient numbers of qualified students and a failure to retain them in the programs once they are there (either because students have difficulty with course work or because business and industry hire them before they graduate). In Japan the quota system in colleges determines how many places are available. Both countries face the problem of technical programs being expensive to operate and run.

PREPARATION AND IN-SERVICE ENHANCEMENT OF TEACHERS AND FACULTY

The initial preparation of teachers and faculty in Japan is similar to preparation in the United States. As with most programs in Japan, one big difference is that almost all students preparing to be teachers in Japan enter colleges and universities knowing from their freshman year that they will be teachers. In the United States, students often do not select teaching as a career until their junior year. One other difference is that the terminal degree in universities does not appear to be as crucial in Japan. At universities, about half of faculty members have doctorates. Most of others have master's degrees; a few, in special cases, have bachelor's degrees. At technical high schools and colleges, teachers have a variety of degrees, although a doctorate or master's is not required. It is anticipated that in the near future school principals will need a master's degree. Most teachers graduate from one of the colleges of education, although a few have engineering degrees and some have general college degrees. As in the United States, school teachers must have a certification to teach certain levels and subjects. Such certification is usually granted by a prefecture in Japan and by the state in the United States. In Japan teachers may have several certifications. Formal education qualifications are less important at the special colleges. In both Japan and the United States, there is an increasing use of part-time teachers at the college level.

While continuing in-service courses are required in the United States for most school teachers, there is little of this in Japan. In Tokyo, technology teachers can take seminars and courses at the Metropolitan Institute, but there does not appear to be any particular incentive for doing so. Salaries are based more on age and length of service than on any additional education and expertise that teachers acquire after graduation. Except for some funds to attend conferences and some national grants to spend time abroad, in-service workshops and short courses for college faculty are not widely encouraged. However, new policies, advocated by the Council on Education encourage better initial teacher training and further in-service education.

ANALYSIS OF SYSTEM EFFECTIVENESS TO DATE

The Japanese system for preparing technicians for an industrial economy based on production has been extremely successful. However, it is based on production, not product development. The Japanese Ministry of Education, Science, Sports, and Culture is concerned about whether a system that has successfully prepared a workforce geared to rapidly producing high-quality products can also become a system that takes the lead in developing a new generation of technical products. It also worries that the number of students entering such programs has dropped drastically over the past 20 years. Companies such as Denso and Toyota report that numbers of students have declined and that they have to recruit harder to secure the number needed to maintain quality.

EMERGING TRENDS, KEY ISSUES, CONCERNS, AND TENSIONS

Encouragement of Creativity and Lifelong Learning

Monbusho reports that the basic philosophy of all education in Japan is now instilling in students a zest for learning and room to group (Ministry of Education, Science, Sports, and Culture 1996a). Policy states that the curriculum must change to encourage more creativity among students. In a country with a national curriculum and a highly specialized system of written examinations that determine the level to which students can go, this is not an easy task. At the national level, there is a movement toward requiring fewer subjects, allowing students more flexibility in choosing courses, and allowing local schools more flexibility. At least at the technical high schools, this may be difficult, because the vocational secondary schools are furnished the same equipment throughout the country. The new system of integrated high schools is encouraging this trend.

Diversity Issues

Men dominate the ranks of faculty, administrators, and industrial personnel in vocational secondary schools and technical colleges. The primary exceptions are English and homemaking teachers. Female students are still not well represented in technical studies at colleges and universities, but their numbers are increasing. For example, in the colleges of technology, about 18 percent of the students are female. In particular, women are entering more programs in computer technology and environmental and science technology areas. There are still very few women students, however, in mechanical, automotive, or electronic technology fields. Women are still underrepresented in the engineering technology fields in the United States also, although there are increasing numbers of women in these fields as well as in the science and technology areas.

Attracting Students to Technical Careers

Both the United States and Japan are finding it harder to attract students to careers as technicians, although companies in both countries show a strong need for highly trained workers. Government, industry, and academia are all concerned about this trend.

Issues for the Future

One basic difference between the approaches to education in Japan and the United States is that in Japan most educational decisions appear to be made for the common good of groups and society. In the United States, decisions are often made to benefit individuals and to encourage individuality and creativity. Yet many issues in the Japanese education of engineering and science technicians are similar to issues in the United States. Among the common concerns are these:

- How can more students be attracted to technical careers?
- How can public colleges and universities provide industry with workers whom industry values?
- How can creativity be encouraged while maintaining a high degree of specialized technical competence?
- How can curriculum changes be incorporated quickly as technology changes so rapidly?
- How can teachers and faculty stay abreast of changing technologies?
- How can secondary schools and colleges provide students with access to expensive modern technology and equipment? How can academic institutions afford to keep equipment and facilities up-to-date to allow students to prepare for the industrialized society?
- What is the proper balance between general education courses and specialized courses?

RELEVANCE FOR POSSIBLE ADAPTATIONS TO THE AMERICAN SYSTEM

There are many lessons that the United States can learn from the Japanese system. Direct adaptation, however, will be difficult since the Japanese system is so clearly a national system. Monbusho can order changes summarily, while similar changes in the U.S. would require as many as 1,500 different, locally controlled community colleges to adopt them. Four lessons of Japanese technical education, however, may be particularly valuable models for the United States:

- Japanese business and industry place a high value on having students complete programs that they begin. Students thus have the benefit of both generalized and special courses. Students also learn that completion of tasks is important.
- The high level of general education provided by the elementary and lower-secondary schools guarantees that students entering vocational programs have the necessary skills to learn. In particular, the level of mathematics and science that all students take is quite high, thus ensuring that students can continue to learn once they are in the workforce.
- There is a high value placed on students understanding the cultures and languages of other industrialized countries. In an age of high international competition, this may prove to be invaluable.
- A system of certification ensures a common level of competence.

ABOUT THE AUTHORS

Elizabeth J. Teles is the lead program director for the Advanced Technological Education (ATE) program in the Division of Undergraduate Education at the National Science Foundation (NSF). The ATE program is designed to improve the education of science and engineering technicians in the United States at both the secondary school and community college levels. Before joining NSF, she taught mathematics at Montgomery College in Takoma Park, Maryland. In 1997, she received a special fellowship from the Japanese Society for the Promotion of Science (JSPS) to visit Japan and to study the Japanese system of technician education.

Hidetoshi Miyakawa is a professor of technology education at Aichi University of Education, Aichi Prefecture, Japan. He is an active member of the International Technology Education Association (ITEA) and executive director of the Japanese Society of Technology Education Association. He has spent considerable time in the United States studying technology education, and translated the ITEA document *Technology for All Americans* into Japanese.

BIBLIOGRAPHY FOR CHAPTER 4

Abe, Yoshiya, ed. 1989. *Non-University Sector of Higher Education in Japan.* Hiroshima, Japan: Research Institute for Higher Education, Hiroshima University.

Aichi University of Education. 1997. *College Bulletin 1997.* Aichi, Japan: Aichi University of Education.

Association of National Colleges of Technology Japan. 1993. *Colleges of Technology in Japan.* Tokyo: Ministry of Education, Science and Culture.

Denso Corporation. 1997a. *Denso Industrial Training Center and Technical College.* Aichi, Japan: Denso Corporation.

———. 1997b. *Denso Technical College Bulletin 1997.* Aichi, Japan: Denso Corporation.

———. 1997c. *Industrial Technology Training Center and Technical College.* Aichi, Japan: Denso Corporation.

Japan Almanac 1997. 1996. Tokyo: Asahi Shinbun Publishing Company. 1996.

Kotaka, Yasukuni. n.d. *An Introduction to Continuing Education at NEC.* Kawasaki, Japan: National Education Center Institute for Technology Education.

Ministry of Education, Science, and Culture. 1994. *Education in Japan: A Graphic Presentation.* Tokyo: Ministry of Education, Science and Culture.

Ministry of Education, Science, Sports, and Culture. 1995. *Japanese Government Policies in Education, Science, Sports and Culture 1995. Remaking Universities: Continuing Reform of Higher Education.* Tokyo: Ministry of Education, Science, Sports and Culture.

———. 1996a. *MONBUSHO.* Tokyo: Ministry of Education, Science, Sports and Culture.

———. 1996b. *The Model for Japanese Education in the Perspective of the 21st Century: First Report by the Central Council for Education.* Tokyo: Ministry of Education, Science, Sports and Culture.

———. 1996c. *Outline of Education in Japan 1997*. Tokyo: Ministry of Education, Science, Sports and Culture.

———.1997a. *Japanese Government Policies in Education, Science, Sports and Culture. 1996 Priorities and Prospects for a Lifelong Learning Society: Increasing Diversification and Sophistication*. Tokyo: Ministry of Education, Science, Sports and Culture.

———. 1997b. *Statistical Abstract of Education, Science, Sports and Culture*. Tokyo: Ministry of Education, Science, Sports and Culture.

Miyakawa, Hidetoshi. 1996. *Curriculum of Technology Education in Japan*. Aichi, Japan: Aichi University of Education.

———. 1996. *School Works of Technology Education in Japan*. Aichi, Japan: Aichi University of Education.

Reischauer, Edwin O. 1988. *Education. The Japanese Today: Change and Continuity*. Cambridge, Mass.: Belknap Press of Harvard University Press.

Tokyo Metropolitan Board of Education. 1997. *Public Education in Tokyo*. Tokyo: Tokyo Metropolitan Board of Education.

Tokyo Metropolitan Institute for the Education of Technology. 1997. Tokyo: Tokyo Metropolitan Institute for the Education of Technology.

Tokyo Polytechnic College Bulletin 1997. 1997. Tokyo: Tokyo Polytechnic College.

Toyota Motor Corporation Human Resources Division. 1997. *Toyota Human Resources Management*. Nagoya, Japan: Toyota Motor Corporation Human Resources Division.

Toyota National College of Technology. 1997. *Toyota National College of Technology College Bulletin 1997*. Nagoya, Japan: Section of General Affairs, Toyota National College of Technology.

University of Tsukuba, Senior High School at Sakado. 1997. Sakado, Japan: University of Tsukuba.

Vocational Education Division of the Elementary and Secondary Education Bureau in the Ministry of Education, Science and Culture. 1994. *An Introduction to Vocational Upper Secondary Schools: Brush Up Your Personality*. Tokyo: The Vocational Education Division of the Elementary and Secondary Education Bureau in the Ministry of Education, Science and Culture.

———. 1994. *What's An Integrated Course?* Tokyo: The Vocational Education Division of the Ministry of Education, Science and Culture.

Wada, H., H. Nagano, K. Sirouzu, T. Mabuchi, and N. Sadan. n.d. *Engineering Education at the Technology Training Center of Mitsubishi Heavy Industries, Ltd.* Yokohama, Japan: Mitsubishi Heavy Industries.

Yoshioka, Yoshio. 1996. Curriculum Reform at Hitachi Ibaraki Technical College for Technical High School Graduates in the Companies of Hitachi Group. Paper presented at International Conference on Engineering Education, April, in Chiba, Japan.

Chapter 5

The Technician Manpower Development System in Israel

Jack L. Waintraub
Middlesex County College, New Jersey
Haya Adner
Queensborough Community College, New York

INTRODUCTION

With a population of less than 6 million, Israel is one of the most technologically developed countries in the world, boasting a high percentage of scientists, engineers, and technologists. Israel places technological power, technological excellence, and technological education at the top of its list of national priorities. According to the director general of Israel's Organization for Rehabilitation by Training (ORT), "the future of the economy, manufacturing and export is no longer detached from the quantity and quality of graduates of the technological education system" (Ben-Ami 1997). Israel's educational philosophy is well rooted in Jewish tradition, which places a strong emphasis on education and study as lifelong obligations. Education is considered to be the key to Israel's national future. Providing scientific and technological knowledge and skills to all students is essential to the country's continuous development.

THE ECONOMY

Israel's economy experienced an annual growth of 6 percent through the 1990s. This phenomenal growth can be attributed to a program of official government encouragement of the high-tech industry and the availability of a highly skilled and well-educated technical workforce. The rapid development of high-tech industries in Israel created a demand for engineers, computer scientists, and highly skilled technicians. In 1996, the combined sales of various sectors of the electronics industry alone had a 10 percent growth from the previous year, shown in Figure 5.1 *(below)*. Israel's export revenues were a strong contributing factor to these growth figures.

In 1996, the electronics industry employed 24,000 engineers, scientists, and technicians—including 14,000 university graduates. Figure 5.2 *(below, right)* depicts the growth distribution of manpower in Israel's electronics industries. These data show a significant percentage of technicians, as compared to engineers and scientists, employed in the electronics industry.

According to the Israel Expert Institute, Israel has the world's greatest number per capita of professionals in the field of science and engineering working on new technologies. Some 140 engineers and scientists out of every 10,000 of its population are employed in research and development. More than 25 percent of the workforce is

Figure 5.1 Israel Electronics Industries Sales

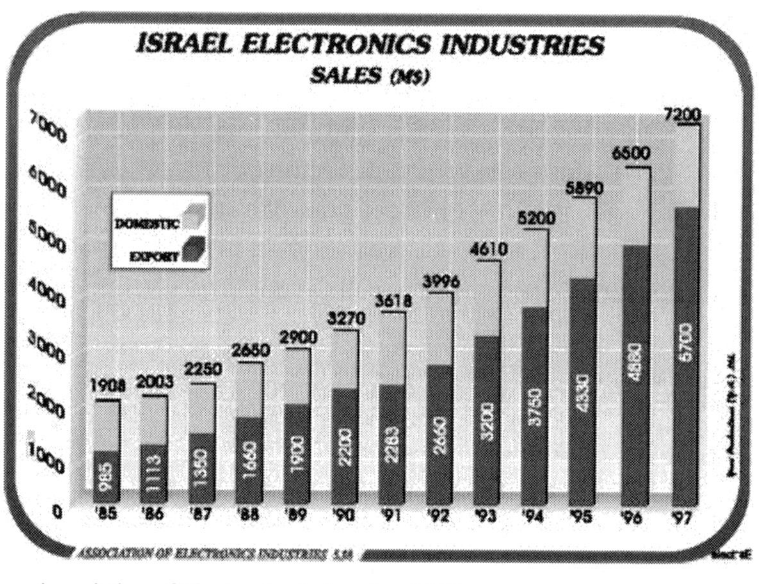

Source: Association of Electronics Industries, 1997

employed in technical professions, as compared to 8 percent in the United States and 12 percent in Japan.

In spite of its fairly high labor costs and its shortage of natural resources, Israel has an advantage over other countries in its region—a highly educated workforce, with 28 percent of the population possessing college degrees in the high-tech arena. The Israeli government has recently spent more than 3 percent of the country's gross domestic product (GDP) per year on programs encouraging high-tech research and development, the highest percentage of all developed countries, as shown in Figure 5.3 *(page 128)*.

Israeli scientists and engineers are the highest per capita producers of scientific publications in the world, and the country has one of the highest patent filing rates per capita. According to Israel's Economic Commission, Israel has a $1 billion software industry.

Israel's compulsory military service, where a large number of young people receive training and experience in technical disciplines,

Figure 5.2 Israel Electronics Industries Employment and Manpower Structure

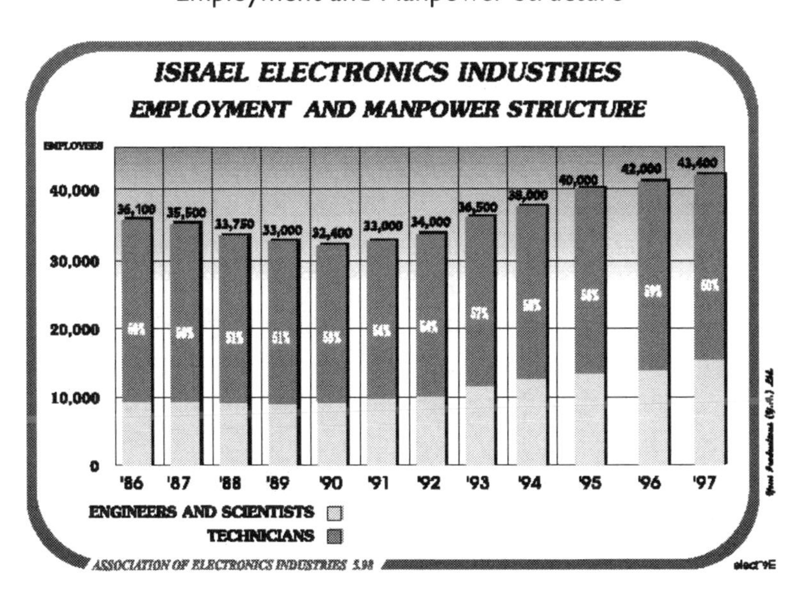

Source: Association of Electronics Industries, 1997

Figure 5.3 Comparison of Israeli and International Investment in Research and Development (Percentage of Gross Domestic Product: 1995)

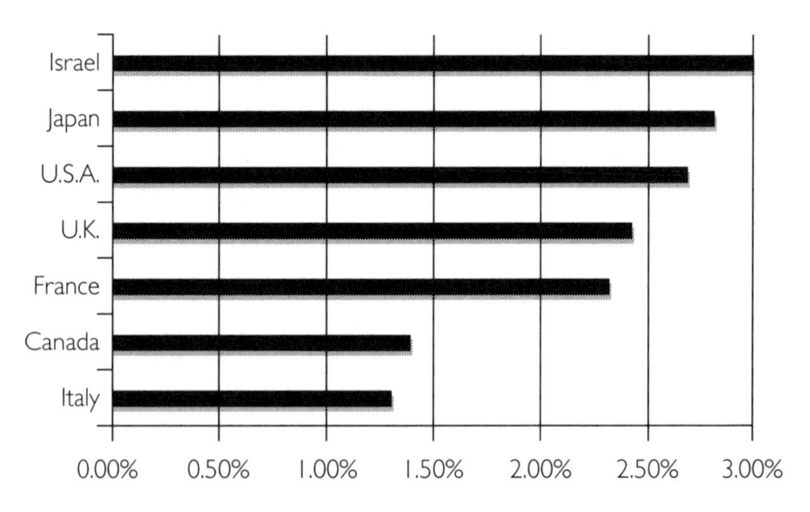

Source: Association of Electronics Industries, 1997

is a strong contributor to the country's workforce development. The military service has proven to be an excellent incubator for nurturing talent for start-up companies in the computer software business, including telecommunications and Internet security. This was pointed out by the economic minister to North America, who said, "When you leave the army at 25 or 26 with a very strong technical background, what do you do? You start a company." (Wang 1997). Not everyone with technical skills starts a company after discharge from military service, but they can become valuable workers for a technical enterprise. No wonder more than 2,000 high-tech companies have facilities in Israel. These include U.S. and other international companies such as Intel, National Semiconductor, and Motorola, which have research and development and semiconductor manufacturing facilities in Israel. Israel is considered an exporter of manufactured products. In 1996, significant industrial growth was experienced in electric motors and equipment, communication, control, medical and scientific equipment, electronic components, and computers.

The demand for engineers and technicians skilled in the high-tech target areas is on the rise. The Industry and Trade Ministry estimated the high-tech labor shortage to be more than 3,000 engineers, with a corresponding shortage of other technical personnel. Recruitment efforts to attract qualified personnel outside of Israel are being conducted. The most acute shortages are in the computer hardware and software areas.

While there is obvious growth of scientific and technical industries, there appears to be a disconnect between industry's demand for technical workers and the output of educational institutions. The shortage of 6,000 computer engineers and software technicians reported in *Link* magazine (Kraus 1995) clearly illustrates the need for a joint effort among educational institutions, industry, and government to increase the support for science and technology education in order to increase the pool of qualified personnel. The same article points out the high interest in technical studies on the part of the students. For example, the Technion (Israel's Institute of Technology) "opened a course for graduates in the industrial sciences who want to learn software engineering. . . . There were 1,500 applicants and only 100 spaces" (Kraus and Fedler 1995). Fear of losing the country's ability to sustain its economic growth potential has spurred a dialogue for involvement in technological education among the leaders of industry, government, and education. Some view the need to supply the national resources required to alleviate the shortage of technically trained personnel as essential to the "development of the state, to its future, and to its security."

THE EDUCATION SYSTEM

The Israeli education system is highly centralized and financed primarily by the national government and local education authorities. Educational policy reflects the need of Israeli society to include social and academic integration through the absorption of immigrants, to promote the advancement of disadvantaged population groups, and to provide equal opportunities in education for all children. The Ministry of Education, Culture and Sport is the oversight authority for education in the country. It is responsible for school curricula and educational standards. The government finances 72 percent of the cost of education; local authorities and other sources make up the rest.

Education in Israel is available from age two, but is mandatory from five to 16 and free up to age 18. Primary school includes studies in grades one through six. Grades seven through nine are intermediate school and grades 10 through 12 are secondary school. There are also postsecondary education, university, and adult education. Postsecondary education includes studies in teachers colleges, technical institutions, and accounting schools. There are four types of schools within the education system dedicated to serving the various cultural groups of the country: state schools, attended by the majority of students with instruction in Hebrew; state religious schools, with emphasis on Jewish studies; Arab and Druze schools, with instruction in Arabic; and private schools.

Secondary Schools

Secondary schools have three different tracks: academic, vocational and technical, and agricultural. Business education is included within the vocational school studies. The majority of secondary schools offer academic curricula leading to a matriculation certificate that is required for entrance to higher education. Matriculation examinations are offered at three levels of difficulty. The level of three units is least demanding and the level of five units is most difficult and most comprehensive. Certain secondary schools provide specialized curricula that lead to a matriculation certificate or a vocational diploma. Technical schools offer programs that lead to a matriculation certificate and preparation for higher education to become technicians with a vocational diploma or to acquire practical skills. Technician education and training is also provided by military preparatory schools that prepare technical personnel in specific fields required by the Israeli Defense Forces (IDF). During the 1996–97 academic year, a total of 433 secondary schools offered general academic programs, compared with 326 that offered vocational and technical programs (Hebrew education). In the same period, there were 97 Arab-education secondary schools that offered general academic programs and 61 schools that offered vocational and technical programs (Central Bureau of Statistics 1998).

Vocational and Technical Education

As with other schools, vocational and technical education is overseen by the Ministry of Education and local authorities. However, a number of nonprofit public organizations manage and finance vocational school networks, such as the Organization for Rehabilitation by Training, by sharing the operational costs with the Ministry of Education and local authorities. The technological tracks continue to be popular with a large segment of the high school population. Graduates of technical tracks are assigned to serve in technical units of the IDF.

Several vocational schools are run by the Ministry of Labor and provide vocational education for students who do not attend regular schools. These students are subject to the Apprenticeship Law. Students enrolled in the apprenticeship programs spend two years in classroom studies followed by one or two years alternating between classroom and work experience. Youth centers combine vocational and special education, enrolling students who do not wish to pursue academic studies. The youth center programs are also work-based, providing practical skills training. A total of 102,268 students were enrolled in vocational training courses in 1996, and another 13,861 participated in supplementary training for adults programs, supervised by the Ministry of Labor. This number includes 1,785 individuals engaged in on-the-job training.

Higher Education

Higher education is conducted at universities and other degree-granting institutions under the authority of the Council for Higher Education (CHE). The council is a public entity headed by the minister of education, academicians, community leaders, and a student. CHE is independent of the Ministry of Education. Higher education is highly subsidized with public funds (70 percent); tuition covers 20 percent and private sources account for 10 percent. More than 150,000 students are enrolled in degree-granting institutions. In Israel, what distinguishes higher education from postsecondary education is that postsecondary institutions provide vocational and professional training, and they do not necessarily require the completion of the state matriculation examination, which is a requirement for entrance to a higher education institution. Many of the technicians and practical engineers in Israel are graduates of postsecondary institutions. In 1996, 14,905 students were enrolled in technician training and practical engineering programs. The higher education system is composed of

- eight universities, including the Open University;
- non-university institutions that provide education at the undergraduate level in specific fields such as teacher training, technology, business administration, and more;
- regional colleges, originally established as satellites of universities to serve students in rural locations, that provide a starting point for their studies toward a degree and allow for completion at the main campus of the university (similar in concept to the community colleges in the United States)
- general colleges that were established with comprehensive course offerings at the undergraduate level.

The latest additions are three institutions strategically located in the northern, southern, and central regions of the country that award a Bachelor of Technology degree with programs in applied engineering. These programs are modeled after the schools of applied engineering in Germany, which successfully train engineers for industry.

The universities perform two functions: teaching and research. Basic research is conducted mainly at the universities. Only seven universities are authorized to offer graduate degree programs. Because of Israel's compulsory military service—three years for men and two years for women—most students begin their studies when they are over 21. However, more than half of Israel's 20-to-24-year-olds are currently enrolled in postsecondary or higher education institutions. A bachelor's degree in universities is earned in most cases after three years of study. Certain programs of study are longer. For example, engineering is a four-year program and architecture requires five years. Some 32 percent of university students are enrolled in computer, engineering, mathematics, and natural sciences programs.

The Open University offers nontraditional education routes toward a bachelor's degree, based on distance learning and, unlike the entrance requirements to the other universities, requires only that applicants be capable of academic study. However, prerequisites for courses of study are listed. The School of Technology of the Open University offers programs leading to technician and practical engineer diplomas. More than 28,000 students are enrolled in courses through the Open University.

Technical Education

Technical education in Israel is multifaceted. In a country where more than 25 percent of the workforce is employed in technical professions, compared with 8 percent in the United States and 12 percent in Japan, technical education is well developed to serve the population at various levels. Science and technology studies are integrated within all levels of education. Students entering secondary school have several choices, but they must meet the entrance requirements of the individual institutions. Some are very selective, especially the premier technical high schools that are affiliated with universities or those that are part of a technical school network. Most students enroll in academic high schools with a curricula concentration in science and humanities leading to a matriculation certificate and preparation for entrance to a university. Some comprehensive high schools provide a broad spectrum of studies that leads to a matriculation certificate or a vocational diploma.

Students' entrance to technical high schools is based on their intermediate school record and aptitude for technical studies. Various paths leading to matriculation and a technical diploma are available to students. For example, students who study at the Tel Aviv University Technical College High School can complete their studies by successfully passing a matriculation examination, presenting a thesis, or teaching a class in one of the technological subjects during their senior year.

Study toward the technician certification and the practical engineer certification can be accomplished by continuing at one of the technical schools or colleges. After two or three semesters of full-time postsecondary studies, students are eligible for technician certification and, with an additional year, for the practical engineering certification. These programs are sometimes referred to as grade 13 and grade 14 studies. Evening part-time studies are also available at these institutions. Evening programs for practical engineering are three years in duration, five evenings per week, year round. The curricula adhere to the requirements of the National Institute of Technological Training (NITT) and are approved by the National Committee for Training of Technicians and Practical Engineers. Secondary school programs operate under the auspices of the Ministry of Education, while grade 13 and 14 programs have joint oversight by the Ministry of Education and Culture, the Ministry of Labor, and NITT. A number of technical colleges are affiliated with universities. Students benefit from university affiliation through sharing of facilities and resources.

Students at the Technion–School for Practical Engineers begin their studies in grade 10 and are prepared to take the matriculation examination upon completion of grade 12. The school emphasizes technological and scientific studies that will prepare students for entrance to the Technion–Israel Institute of Technology. The institution offers programs in a number of technical disciplines, from electrical power and controls, through CAD/CAM, to computer technology. Those graduating from grade 12 in the electrical power program are licensed to work as electrician helpers. Similar certifications are afforded to graduates of other programs. Students who pass the matriculation examination at a certain level (five units, which is the highest level of achievement) in mathematics, physics, and technical studies receive preference for entrance to the Technion. Special consideration for service in appropriate branches of the military is given to graduates of technical programs, providing opportunities for further experience in the field. Continuation of studies for an additional year (grade13) makes one eligible to take the examination to become a certified technician. An additional year of studies (grade14) prepares graduates for the examination for certified practical engineer.

Requirements for certification include passing the national examination and completing of an oral and written defense of a comprehensive project before a team of examiners chosen by the National Institute for Technological Training. The project is a capstone experience of an authentic industry task, which might require the design or development of a product or a solution to a problem performed in an industrial setting with professional advisement. Those completing grade 14 studies are eligible for admission to the new applied engineering programs leading to a bachelor of technology (B.Tech.) degree. Upon evaluation of grades and accomplishments toward the practical engineer certification, advanced standing can also be given toward an engineering program. Certain branches of the military service offer select graduates of grade 14 scholarships toward B.Tech. studies before they enter the service and the possibility of an officer track.

Institutions that offer postsecondary technical programs provide preparatory courses and remediation in specialty subjects. Day and evening preparatory programs are available. Entrance requirements vary from institution to institution; however, the program of study is prescribed by NITT. The entrance requirements to the Tel-Aviv

University College of Practical Engineering (TAU-PE) are matriculation with three units each of mathematics, physics, and English, and one unit of Hebrew. Graduates of programs with a certified technician diploma start in the second year. The college also offers the following programs of study: retraining of engineers to become teachers of technological professions; update courses in technological professions for engineers, especially new immigrants; and training of technicians and practical engineers for military related jobs.

Half of the population of technicians and practical engineers in Israel gained their education at one of the Organization for Rehabilitation by Training schools. ORT schools provide secondary vocational and technical education, as well as postsecondary technician and practical engineering programs. ORT also pioneered the offering of academic programs in technology that lead to a Bachelor of Technology degree and a Bachelor of Technology Education (B.Ed.Tech.) degree for teaching careers in science and technology.

The Pedagogy Center for Research and Development of ORT is dedicated to the professional training of the teaching personnel in the network. The center's work includes

- designing and developing curricula;
- publishing instructional materials jointly with the Ministry of Education;
- providing institutions with educational support in the use of technology by students and teachers;
- consulting in the development of technological centers for training teachers and in the adoption of curricula;
- developing multimedia instructional materials.

The continuing education division of ORT, called Career ORT, provides flexible programs that include workforce training and retraining in a variety of vocational and technical areas, as well as preparatory courses for matriculation, retraining for immigrants to the country, and various programs for business and industry. Some 20,000 students are enrolled in Career ORT annually.

The postsecondary technology curricula for technicians and practical engineers program prescribed by the National Institute for Technological Training are basically similar at different institutions. The practical engineering program is 2,560 hours in duration and can be completed in two years—four semesters on a full-time basis. A semester lasts for 16 weeks, with an average of 40 hours of study per week. Some institutions are on a trimester basis with a equivalent total hour distribution. (Students enrolled in engineering technology programs in the United States study an average of 24 hours per week and a semester averages 14 to 16 weeks.) Table 5.1 *(page 138)* shows a sample practical engineering program with an hourly breakdown per semester.

The technician program can be completed at the end of year one. In order to graduate and receive certification, students take internal and external (national) examinations and defend their final project.

The new Bachelor of Technology programs, established in response to the demand for personnel in Israel's high-tech industries, are offered under the auspices of the Council for Higher Education. They offer a four-year engineering curriculum, building on the applied nature of the two-year practical engineering programs. Students are admitted as freshmen or with a practical engineering diploma into the third year of studies after completing supplementary courses.

The School of Technology of the Open University of Israel is a nationwide technological institution that prepares individuals for two certification diplomas: technician and practical engineer. The school also offers a wide variety of refresher courses tailored to the needs of industry, and vocational training programs for certification and licensing. Education is provided through some 20 courses that are modular, offered on a semester basis, and self-contained. The delivery method includes self-study written material, home labs, computer software and simulation programs, computer communication for distance learning, project work, tutored group sessions at study centers, various home assignments, and final examinations in every course. The school offers three technological programs: electronics, software

Table 5.1 Typical Electronics Technology Program in Israel: Hours per Semester

	YEAR 1		YEAR 2		
Subject	**Semester 1**	**Semester 2**	**Semester 3**	**Semester 4**	**Total Hours***
GENERAL					
Mathematics (theory & application)	7	6	–	–	208
Physics	5	3	–	–	128
Technical English	3	2	–	–	80
Drafting	–	2	–	–	32
FOUNDATION					
Electricity	7	6	–	–	208
Computer Programming	–	3	–	–	48
Digital Systems	5	3	2	–	160
Introduction to Computers	2	–	–	–	32
Analog Electronics	–	6	5	6	272
Digital Electronics	–	–	4	4	128
Measurements	–	–	3	–	48
Measurements Power Systems	–	–	2	–	32
COMPLEMENTARY					
Introduction to Communications	–	–	3	–	48
Introduction to Controls	–	–	2	–	32
Introduction to Electro-optics	3	–	–	–	48
SUPPORT					
Digital Computers and Microprocessors	–	–	6	5	176

	YEAR 1		YEAR 2		
Subject	**Semester 1**	**Semester 2**	**Semester 3**	**Semester 4**	**Total Hours***
Programming Systems	–	–	–	3	48
Computer Networks	–	–	2	2	64
ELECTIVE	2	–	–	–	32
LABORATORIES					
Electricity and Measurements	3	2	–	–	80
Computer Programming	–	2	–	–	32
Digital Systems	–	3	–	–	48
Introduction to Computers	2	–	–	–	32
Analog Electronics	–	2	3	3	128
Digital Electronics	–	–	3	4	112
Digital Computers and Microprocessors	–	–	4	4	128
Microprocessors Software Systems	–	–	–	3	48
Computer-Aided Design and Imaging	–	–	1	2	48
PROJECT LABORATORY	–	–	–	5	80
Total Hours Weekly	39	40	40	41	2,560

*Total hours are obtained by multiplying weekly hours by 16 weeks per semester.

engineering, and industrial management. The technological courses were developed by the Center for Educational Technology.

Because of the clearly defined educational lines by educational preparation and specific certification among a technical high school diploma, certified technician, certified practical engineer, and the new Bachelor of Technology degree, industry is well prepared to employ technical personnel at the needed level. Well-established skills standards provide for a better match of an individual's technical skills with required job tasks.

A number of institutions also exist for the purpose of training and retraining professionals in the high-tech field. For example, the Hi-Tech College of the Center for Technology Studies is specifically targeted at the development, engineering, manufacturing, and management sectors of the Israeli high-tech community (Hi-Tech College 1998). The training programs are specifically designed for employed professionals who need technical updating on a given subject or to learn a new subject. Training programs with certification are product- or company-specific.

Adult education is available in an ever-increasing range of courses and programs. The Ministry of Education, as well as private and public institutions, offers courses of varied interest to the public. The Ministry of Labor, in partnership with industry, offers vocational training and retraining programs, in both day and evening classes. Courses are taught at special centers, as well as in institutions for technological and professional training. Retraining programs are also offered to a large number of immigrants with higher education degrees in professions no longer in demand.

CONCLUSION

Israel's robust economy, as a result of a strong global market for its technology exports, has triggered an ever-growing demand for a technically skilled workforce. Having a well-developed technical educational system for multiple skill levels and a well-established national certification system helps provide qualified electronic technicians and practical engineers for the high-performance workforce. The government's support and control of the educational process has served to establish rigorous curricula and maintain the quality of graduates.

Opportunities for technical education are plentiful, beginning at the secondary school level. Students are prepared for the trades, for postsecondary technical education, or for entry to the university. Postsecondary education prepares students for national technician or practical engineering certification. Graduates of practical engineering programs now have opportunities to continue toward a Bachelor of Technology, the first academic degree. Articulation paths among the various levels are clearly defined.

Israel established itself as a powerhouse of technological and scientific research and development due to its enormous human talent. It is also a manufacturer of high-tech software and hardware in a multitude of industries. Shortages of skilled technical personnel in the information technologies, microelectronics, biotechnology, and other fields are forcing companies to recruit personnel from other countries. According to educational and industry leaders, the gap between demand and supply of technical and scientific personnel can be bridged through collaboration among academia, industry, and government. A factor that is often pointed to in discussions about increasing the pool of technical workers is the poor marketing of technical and scientific careers, in spite of the strong demand for these professions. All the stakeholders have begun discussions on how to improve the situation. These sentiments are expressed by educators and industry leaders who would like to see a more direct marketing approach.

Service in Israel's defense forces provides a fertile base for training of technical personnel and for cultivating future entrepreneurs, especially in the information technology field. According to one report, "the seeds of nearly every hi-tech start-up are sown in the military where Israeli youths bond with their fellows and learn to innovate and improvise" (Tuck 1997).

With a strong demand for technical personnel and a well-defined vocational and technical educational system that targets a broad segment of the population, Israel will continue to have a distinct edge in a competitive global environment.

ABOUT THE AUTHORS

Jack L. Waintraub is professor and chairman of the Physics/Electrical Engineering Technology Department at Middlesex County College and director of the New Jersey Center for Advanced Technological Education. He is a licensed professional engineer and the author of several textbooks in electrical engineering technology. He served as a program director at the National Science Foundation during 1993–94 and currently serves as a vice chair of the Technology Accreditation Commission of the Accreditation Board for Engineering and Technology.

Haya Adner is associate professor of mathematics and computer science at Queensborough Community College of the City University of New York. Holder of two degrees from the Technion, Israel Institute of Technology, she is the author of several articles on mathematics education. Her interest is in mathematics modeling, technology, and interdisciplinary education.

BIBLIOGRAPHY FOR CHAPTER 5

Ben-Ami, Haim. 1997. *Technological Education: Excelsior!* Jerusalem, Israel: Organization for Rehabilitation by Training: Colleges and Schools for Advanced Technologies and Sciences.

Central Bureau of Statistics. 1998. *Schools, Teaching Posts and Pupils: Schools in the Educational System.* Jerusalem, Israel: State of Israel Central Bureau of Statistics.

Hi-Tech College. 1998. *The Hi-Tech College Story*. Herzelia, Israel: The Hi-Tech College.

Israel Ministry of Foreign Affairs. 1997. *Education: Higher Education.* Jerusalem, Israel: Israel Ministry of Foreign Affairs.

Kraus, Orli, and Jonathan Fedler. 1995. "Wanted: More Engineers Now." *Link*, November-December.

Sandler, Neal. 1997. "Israel Taps Jewish Communities for High-Tech Immigrants." *Tech Web*, January 20.

Tuck, Barbara. 1997. "Israel: A Remarkable Hotbed for Hi-Tech." *Computer Design*, July 16.

Wang, Nelson. 1997. "How Tiny Israel Became an Internet-Industry Giant." *Web Week*, January 20.

Chapter 6

The German Vocational, Apprenticeship, and Continuing Education System

Ashok Agrawal
St. Louis Community College
at Florissant Valley, St. Louis, Missouri

HIGHER EDUCATION SYSTEM

German higher education institutions are rooted in the European university tradition. The great majority of higher education institutions in Germany are state institutions maintained by the Länder (federal states). There are currently more than 300 institutions, including 91 universities, 7 comprehensive universities, 11 independent colleges of education, 19 theological colleges, 43 colleges of arts and music, and 153 *Fachhochschulen*. The total also includes 62 privately maintained institutions.

The university-level higher education institutions are the largest part of the higher education system, offering courses in law, economics, humanities, social sciences, medicine, natural sciences, and engineering. Several institutions specialize in one branch of study (such as theology, medicine, veterinary medicine, public administration, and sports), and some colleges focus on education to the extent that the training of teachers has not been incorporated into the universities.

Fachhochschulen

A more recent type of higher education institution is the Fachhochschule. At the beginning of the 1970s, it emerged from various institutions of advanced practice–related vocational education, notably the schools of engineering and higher technical schools. Fachhochschulen offer more practice-related three- or four-year study courses of a scientific or artistic nature, especially in the fields of technology, agriculture, economics, design, social work, social pedagogy, public administration, and paralegal services.

By establishing Fachhochschulen, the Länder took account of increasing demands in occupations—resulting from scientific and technological progress—that had created qualitatively new requirements in education. For example, engineering studies oriented to these new requirements had to provide significantly greater methodological and scientific depth, and existing education facilities were unable to offer such an improvement. The growing demand for education, especially the increasing demand for higher education leading directly to occupational qualification, was also an important factor in the creation of these institutions.

The educational tasks and characteristics of Fachhochschulen within the German higher education system can be summarized as follows:

- close links between science and practice in both instruction and studies;
- efficient organization of studies and examinations;
- shorter periods of study (compared to other types of higher education institutions);
- applied research and development.

Professors hired by Fachhochschulen are expected to have specific qualifications, including having completed studies at a higher education institution, together with proof of special capability for scientific work (as a rule, a doctoral degree) or for artistic work. In addition, professors must have suitable pedagogical qualifications and at least five years of professional experience, of which at least three years must have been gained outside the higher education sector. Professors at Fachhochschulen are currently required to teach 18 weekly hours per semester. Most Länder permit Fachhochschulen professors who are active in research to reduce their teaching load. In some Länder, Fachhochschulen professors also may, at regular four-year intervals, take a six-month leave of absence from their teaching duties in order to strengthen their knowledge of current practice in their technical areas.

Routes to Higher Education

There are several routes to higher education in Germany. More than 70 percent of new students at higher education institutions have graduated from a *Gymnasium* (grammar school). Holders of the Abitur certificate awarded by a Gymnasium upon completion of 13 years of schooling are entitled to study the subject of their choice at a university or equivalent institution. Holders of the Abitur awarded by a *Fachgymnasium* (specialized grammar school) are entitled to commence higher education studies only in the field in which the Fachgymnasium specializes. Fachgymnasium graduates earn the Fachgebundene Hochschulreife, a subject-restricted higher education entrance qualification.

The certificate of Fachhochschulreife is required for study at a Fachhochschule. It is usually obtained upon successful completion of the 12th and final year at a *Fachoberschule* (vocationally oriented upper secondary school). Increasingly, however, young people holding an Abitur certificate are enrolling at Fachhochschulen. In 1991 just under half the new entrants at Fachhochschulen held a university entrance qualification. Persons who did not attend or did not complete Gymansium, or who are already employed, may enter schools or *Kollegs* (institutions of full-time adult education), or may enter by passing an examination for the gifted (second educational route).

Access

Germany has made a strong effort to make higher education accessible to all, including those children who, due to their families' occupation, education, or income, formerly had little chance of obtaining a higher education. The availability of student aid has played an important role in this development. At the beginning of the 1950s, only 6 percent of the children born in a given year commenced studies at a higher education institution. Forty years later, over 30 percent did so. The number of students enrolled at universities and Fachhochschulen reached an all-time high (1,875,000) in the 1993–94 winter semester. In 1996, this number was 1,838,000.

In order to facilitate equity of opportunity in education, the state provides students with the funds necessary to cover their educational and living expenses. The amount of financial aid varies according to the incomes of the student and his or her parents and spouse. Housing requirements also affect the funding support level. Until recently, the maximum entitlement for students not living with their parents was DM 940 per month in the old German Länder and DM 855 in the new Länder. Half of the amount paid to each student was provided as a grant, and the other half as an interest-free loan.

Adult and Continuing Education

Adult education and training is provided by the constitution of the Federal Republic of Germany. Germany Basic Law mandates freedom of religion, of conviction, and of speech, as well as the right to assemble and to choose one's occupation. It also supports the right of every person to personal growth.

The terms "adult education" and "continuing education" are used in Germany to refer to continuing adult learning. Until the 1960s, adult education referred primarily to open, general education, aimed especially at cultural and political personal development. Continuing education has been the accepted term since about 1970 for the continuation or resumption of organized learning following the completion of a first educational phase. This term refers primarily to systematically planned, professionally organized educational programs, especially those offered for further vocational qualification. The two terms are now sometimes used synonymously and sometimes used to refer to the two different emphases.

It is still customary to divide continuing education into the two main areas of general and vocational further education. Political and cultural further education is normally classified with general continuing education. Development of general key basic qualifications—such as creativity, communications skills, ability to work in a team, initiative, and constructive computer skills—has also acquired increasing importance in recent years in Germany. Today, it is virtually impossible to distinguish between general and vocational further education in the area of "key qualifications." The largest area of continuing education in Germany—the study of foreign languages and cultures—can be just as important in an occupational context of export and product modification for other countries as it can be in the context of personal growth. Interest has been increasing recently in forms of on-the-job learning that are not regularly organized, and in informal self-directed learning within a social context.

According to the last "Continuing Education Reporting System" (Dohmen 1997), in 1994, some 42 percent of Germans between the ages of 19 and 64 participated in organized continuing education. This represents a 5 percent growth compared to 1991 figures. In 1994, 20 million adults (out of a population of 81 million) took part in continuing education courses, lectures, or study courses. Some 26 percent of these adults attended general-education events, and 24 percent attended vocationally oriented further training events. Fifty-two percent of the participants pursued general further education, and 44 percent attended instruction in vocational further training.

In 1994, the most popular programs were foreign languages and health education, along with commercial or business courses and computer skills learning. The strongest motive shared by all participants for participating in continuing education was a desire to be able to cope better with daily life. The age group most frequently represented in general continuing education courses was 19 to 34; in vocational further training, it was 34 to 49. Sixty percent of persons with the Abitur, the highest secondary school qualification, took part in continuing education events. The corresponding figure for the group of those with a medium-level qualification (on the Realschule secondary school level) was 47 percent. Only 29 percent of adults with a lower school qualification took part in continuing education events.

While 50 percent of employed people took part in continuing education, only 29 percent of the unemployed did so. Forty-one percent of skilled, specialized workers, but only 27 percent of unskilled and semiskilled workers, participated in continuing education.

Many different organizations provide continuing education:

Community adult education centers. These *Volkshochschulen* (VHS) are educational centers without lodging facilities. They concentrate exclusively on continuing education and offer a comprehensive basic curriculum. As a rule, the VHS are community (in cities and rural districts) further education centers. The communities in which they are located normally receive Länder subsidies. The Volkshochschulen, with 28 percent of all participants in general further education, make up the largest segment of this area of training.

Corporations. Companies, both private and public, are the most important providers of vocational further training. They serve 53 percent of all participants in vocational further training, either in their own company training facilities or through cooperation with intercompany further training institutions and other providers.

Private trainers. Private commercial training institutions are especially active in the area of vocational retraining, acquisition and expansion of vocational qualifications, computer training, foreign language instruction, and course work to obtain formal training qualifications. In 1994, they accounted for some 9 percent of all participants in these activities.

Higher education. Institutions of higher education and academic and scientific societies offer scientific and specialized further education. These institutions account for approximately 8 percent of continuing education participants.

Intermediate organizations. The Chambers of Industry and Commerce, Chambers of Craft and Trades, Chambers of Agriculture, and professional associations offer an inter-company framework for vocational training for purposes of adjustment and promotion, and they administer tests for recognized qualifications.

Unions. In the further training sector, unions concentrate primarily on helping members to participate actively in political life, and offer broad political and vocational education.

Community and other public groups. Churches, charity organizations, correspondence schools, residential adult education centers, libraries, and public broadcasting networks also provide further training on a lesser level than the organizations noted above.

Publicly subsidized continuing education institutions are all obligated to accept all interested applicants. Germany's Basic Law 2.1 provides all persons the right to personal development opportunities as they choose. The law implies a public responsibility to provide public funding in order to enable those who wish to participate in adult education or further training to do so.

Total direct expenditures for continuing education in Germany in 1994 was roughly estimated to be more than DM 70 billion. Over half of this funding came from private sources (DM 27 billion from companies and DM 10 billion from the participants themselves). Public funding is primarily from unemployment insurance (DM 13.5 billion) and from the budgets of the Länder, the communities, and the federal government (a total of DM 10 billion).

Vocational Training

Vocational training to become a skilled blue- or white-collar worker is conducted primarily within what is known as the dual system of vocational training. The dual system is a combination of learning in the "serious" world of a company career and learning in the vocational college. The companies concentrate on practical knowledge, while the vocational colleges concentrate on theory. The dual system has been in existence since the 19th century, when the further vocational training colleges were developed. These colleges were the predecessors of today's vocational colleges and were attended by apprentices. The current system of vocational training received its name from the German Committee for Education and Training (1953–1965), which coined the term "dual" for training in a company and in a vocational college in its 1994 *Expert Report on Vocational Training and Education* (Rolf 1996).

The Vocational Training Act of 1969 was a milestone in the development of the dual system of vocational training. Up to the time of its passage, vocational training in the dual system was regulated only in some sectors. The act now presents an extensive and detailed regulatory work that emphasizes, promotes and safeguards public responsibility in vocational training. Also, the act granted those social groups that are particularly interested in vocational training—namely employers and employees—a decisive role in managing the system. The act required the appointment of employer, employee and vocational college representatives to the committees and bodies that organize and oversee vocational training. The act also legally sanctioned the dual system of vocational training as the predominant qualification system in Germany.

Although most of the apprentices in the dual system are trained in private sector companies, the state has a strong public interest in the quality and content of training. The state is concerned that the training meet the specific needs of the sponsoring company and at the same time provide support for the trainees' broad career goals. In-plant training is considered not only a private company task but also a social task, a matter of public responsibility.

Therefore, two principles that are not always in harmony dictate the dual system:

- On one hand, each company may generally decide if, how many, and which young people it wants to train, and which is the best way for it to provide its apprentices with the knowledge that is necessary for a particular profession.

- On the other hand, the state sets up a framework by laws and decrees in order to make sure that vocational training in the dual system comes up to the expectations of both the employment system and society.

The Vocational Training Act of 1969 provides that young persons up to 18 years of age may be trained only in a recognized training occupation. The detailed course of training is specified in the appropriate training regulations, ensuring that training is conducted to uniform standards throughout the country.

Intermediate organizations function between the state and companies. These organizations translate law and regulations into practice. The Chambers of Industry and Commerce, the Chambers of Craft and Trades, the Chambers of Agriculture, and the professional associations (as responsible authorities) ascertain whether an employer has the necessary qualifications to provide training according to state regulations. The training period ends with a final examination, conducted by the chambers, to prove that the candidates have successfully completed their training and possess the required qualifications as skilled workers. A board of examiners consisting of at least three members administers the examination. Employers and employees are equally represented on the board, and at least one member must be a vocational school teacher with an advising vote.

The most important regulatory components of the dual system are

State regulatory component. The state determines the legal parameters of vocational education by means of law and regulations on training.

Market economic regulatory component. The firms supplying training offer training places on the training job market; they decide on the details of the training contracts and carry out the training in keeping with the legal guidelines.

Delegation of regulatory competence to self-organized agencies in the economy. The Chambers of Industry and Commerce oversee the legal and regulatory norms for vocational education in their capacity as public autonomous agencies.

Corporate regulatory component. At all levels of this system, representatives of the employers' and employees' associations participate on an equal footing as policymakers.

Since the Vocational Training Act gave trade unions the right to representation on all federal and Länder committees that regulate vocational training, the unions and employers' associations are, officially, equal partners in national vocational training policy. The self-regulation of organizations involved in vocational training relieves the state of the costs and risks that would be inherent in its own intervention. The participation of the social partners in the regulatory process results in an effective management of the dual system of vocational training, and also helps ensure that the different levels of performance (individual sectors or training groups of training enterprises) are considered when complying with acceptable minimum standards for training.

In the late 1980s, more than two-thirds of the young people born in one age class completed a traineeship (apprenticeship) in one of the recognized training occupations. School-based training programs play a relatively minor role in numerical terms. Most of the apprentices in the dual system are trained in private sector companies. A contract between the training enterprise (the firm where the practical training takes place) and the trainee forms the basis of training under the dual system. Trainees spend three to four days a week at their training enterprise and attend a state-run, part-time vocational school on the remaining days. Blocks of training at external training centers often supplement on-the-job-training. These are generally run by the Chambers of Industry and Commerce or Chambers of Handicrafts, and are mainly attended by trainees who are learning their trade in small enterprises.

About half of all training contracts are concluded with companies that employ fewer than 50 persons. Less than half of all small enterprises provide vocational training, however. Generally, enterprises finance their own training. The operating costs of external training centers are covered mainly by the membership fees that enterprises pay the district chambers and by the attendance fees charged for courses. The federal and local governments finance part-time vocational schools.

Delegation of regulatory tasks to trade associations is most prevalent in areas where the state decrees that consensus between employers' associations and trade unions is a prerequisite for public regulation, as in the case of training regulations for officially recognized apprenticeship trades and occupations. Ever since the enactment of the Vocational Training Act, these apprentice trades and occupations are in-plant training courses regulated by federal ordinances. In contrast, curricula for part-time vocational schools are set out in accordance with the educational statutes of the federal Länder. The individual Länder cooperate on a voluntary basis in the core curricula committees of the Standing Conference of Ministers of Education and Cultural Affairs. The results are of a self-binding nature for the Länder. Although the federal training regulations (related to in-plant training) and core curricula (related to part-time vocational school) are drafted separately, there is coordination in the preparatory phases.

There are currently about 380 recognized skilled trades and occupations that require specialized training. Each of these has a special training regulation that is set by the federal minister for education and science in accordance with the employers' associations and trade unions. These regulations are based on the 1969 Vocational Training Act. They must specify at least the following:

- the name of the trainee occupation;
- the period of training, which shall not normally be more than three or less than two years;
- the skills and knowledge to be imparted in the course of training (occupational description);
- an outline on the syllabus and timetable (overall training plan);
- the examination standards.

The Vocational Training Act requires that training programs focus on a wide range of basic vocational skills, as well as on knowledge of specific occupation information. The dual system offers both training for the special needs of a single company, and a pattern of marketable vocational qualifications according to the requirements of a vocational structured employment system. The final vocational certificates are meant to offer young workers greater prospects of mobility.

The challenge of new technologies and corresponding industrial processes and the varying speed and intensity of their application in different kinds of businesses have made new demands on the skilled workforce. The big industries in particular expect their workforce to be able to promptly apply its specialist expertise on the one hand and to offer adaptations to changing applications with different requirements on the other. Skilled workers are expected to carry out their work independently and flexibly, taking personal responsibility for the quality of their products. Therefore, the training of young people to meet these new challenges has become a priority.

New training regulations, such as those for the industrial metal occupations, have been developed to form the legal framework for these new training courses. Until 1987, there were about 40 metal-working occupations. Now there are only six internally differentiated metal occupations. All of these new courses last for three-and-a-half years. Training begins with one year of basic vocational training, which covers the widest range of skills and knowledge for metal-working. This is followed by specialized training related to occupation groups, (industrial mechanic, tool mechanic, etc.) and then by 16 areas of specialization (within the cutting mechanic occupation, for example, areas of specialization include turning technology, automatic lathe technology, and milling and grinding technology).

FEDERAL INSTITUTE FOR VOCATIONAL TRAINING: BIBB

Bundesinstituts fur Berufsbildung (BIBB) is the key federal organization that ensures the quality of vocational training in the country. It is responsible for approving the standards developed by the representatives of the social partners. It also provides the leadership, structure, and statistics for the apprenticeship program. Apprentice examinations are given by the Chambers of Commerce, which are organizations of employers in each industry. BIBB's interests and activities include the following:

- training trainers and assessing the use of technology (including multimedia technology) for training;
- supporting pilot programs and funding pedagogical research to improve learning and teaching;
- funding enterprises (industries) to look at innovations (the average project costs $1.5 M for three years);
- using universities to consult on and manage innovative projects with enterprises;
- comparing the strengths and weaknesses, benchmark data, and other aspects of Germany's dual system with those of other countries;
- investigating how the United States and other countries assess outcomes and use technology to deliver education.

Structural changes in the workplace and in the economy have generated concerns at BIBB. Among them:

- Students are moving away from the dual system because of their perception that it is a dead-end track with no career mobility. Vocational education must be made more attractive. Twenty-five years ago when the dual system started, 60 percent of students were in Hauptschule and only 10 percent in Gymnasium. Today about 30 percent prefer to go to Gymnasium and only 25 percent to Hauptschule.

- The levels of skills can vary significantly in the dual system program, such as the levels among bank workers, high-end electronics workers, and low-end craftsmen.

- It is difficult to anticipate or identify new occupational fields for the dual system.

- Trade unions have lost power since the recent recessions. Under the apprenticeship system, labor unions must have equal power with employers if they are going to be strong social partners.

SUMMARY OF OBSERVATIONS

General

- Germany's famed dual system consists of a combination of formal schooling and work experience. The work experience is highly structured and is institutionalized throughout the country.

- There is a strong indication that the five technical job categories of the past (research scientists and engineers, practicing engineers, technicians, skilled workers, and unskilled workers) will be compressed into three categories (research scientists and engineers, engineers and technicians, and skilled workers).

- The present workforce is populated with workers of whom 70 percent have gone through the apprenticeship program. However, over the past decade there has been a shift of students going to the university track rather than through the apprenticeship program.

- Industry complains that university trained engineers are good theoreticians but are weak on practical applications. Industry prefers the Fachhochschule engineer who has had several years of mandatory industry or apprentice experience. The trend appears to be toward hiring highly trained apprentices (craftsmen) and graduate engineers of Fachhochschule.

- Skilled workers have high esteem and are well respected in the society. However, it appears that this may be changing.

- The requirement for one year of military or civil service before entering the workforce appears to have a positive effect on the maturity and perspective of students.

- Comprehensive outcomes assessment appears to be the key to maintaining and improving the level of competency of all graduates.

- The dual system is successful in providing a very high level of skill among all workers, but its system for setting standards for new occupational training is very slow. In some cases, it can take 10 years to upgrade an occupation. It does not respond well to the need to develop workers for new or rapidly changing occupations.

- Because employers pay the cost of skill training, they do not encourage higher skill training unless it is necessary. In contrast, the trade unions lobby for broad and high-level skill training that is transferable.

Secondary School System and Universities

- Germany offers a highly structured three-track system of secondary education for preparing students for university, college or technical university, and skilled jobs. Students can move from one path to another based on their performance.

- Assessment and placement activities at the end of the sixth grade determine which of the three types of schools students will attend. Parent-teacher dialogue is key to the decision-making process.

- Schools are not funded by property taxes, but by Länder and federal taxes. Rich states support poor states.

- Everyone must attend school until age 18. Education up to the college level is free. Private schools are also supervised by the state. The school year is longer than in the United States, with only six weeks of summer vacation.

- Secondary teachers are required to have a degree in the major field they teach, and must be experts in at least two subject areas. Teacher salaries are higher than in the United States.

- Curricula are set at the federal level, but individual teachers decide what textbooks and teaching materials are used. Secondary teachers have significant latitude in text selection and classroom activities.

- Technical teachers have the option to spend up to six weeks in industry at full pay. However, few do. School must provide substitutes for teachers who take this option. All teachers are required to take courses in pedagogy.

- In a technical university, faculty must have industrial experience to attain the highest professorial rank.

- Final (graduation) examinations are proposed by colleges and approved by the state.

- Students pursuing university track education may take as many technical and mathematics courses as they desire.

Industrial Commitment

- Industry views its involvement in and support of the German dual system apprenticeship program as a social and moral commitment. Industry bears all expenses and provides facilities, instructors, and industry mentors for the program.

- One steel company is spending more than 3 percent of its payroll on apprenticeship training. In 1991, companies spent a total of DM 42.3 billion on training, equaling approximately DM 30,000 a year for each apprentice.

ABOUT THE AUTHOR

Ashok Agrawal is chair of the districtwide Engineering and Technology Department, St. Louis Community College at Florissant Valley, Missouri. Before joining the college, he was associate professor and chair of the Department of

Engineering Technology at West Virginia Institute of Technology. He is a registered professional engineer in West Virginia. Agrawal is active in professional associations and in 1996 received the Fredrick J. Berger Award for Excellence in Engineering Technology from the American Society for Engineering Education. Through fellowships, he has studied technical school systems in Germany, Japan, and Mexico.

BIBLIOGRAPHY FOR CHAPTER 6

Federal Ministry of Education, Science, Research and Technology. 1994. *Framework Act for Higher Education*. Bonn: Federal Ministry of Education, Science, Research and Technology, Public Research Division, D-53170.

———. 1996a. *The Fachhochschulen in Germany*. Bonn: Federal Ministry of Education, Science, Research and Technology, Public Research Division, D-53170.

———. 1996b. *The Federal Government's Tasks in Education, Science, Research and Technology*. Bonn: Federal Ministry of Education, Science, Research and Technology, Public Research Division, D-53170.

———. 1996c. *Higher Education in Germany*. Bonn: Federal Ministry of Education, Science, Research and Technology, Public Research Division, D-53170.

———. 1996d. *Vocational Training in Germany*. Bonn: Federal Ministry of Education, Science, Research and Technology, Public Research Division, D-53170.

———. 1997a. *Basic and Structural Data 1997/98*. Bonn: Federal Ministry of Education, Science, Research and Technology, Public Research Division, D-53170.

———. 1997b. *Vocational Training in the Dual System*. Bonn: Federal Ministry of Education, Science, Research and Technology, Public Research Division, D-53170.

———. 1997c. *Vocational Training Promotion Act*. Bonn: Federal Ministry of Education, Science, Research and Technology, Public Research Division, D-53170.

Gunter, Kutscha. n.d. "General and Vocational Education and Training in Germany-Continuity amid Change and the Need for Radical Modernization." www.uni-duisburg.de/FB2/BERU/publikat/young.

Gunther, Dohmen, and A. Tubingen. 1997. *Continuing Education in Germany*. Bonn: Federal Ministry of Education, Science, Research and Technology, Public Research Division, D-53170.

Indermit, Gill, and Dar Amit. 1996. *Germany's Dual System: Lessons for Low and Middle Income Countries. Constraints and Innovations in Reform of VET*. Washington D.C.: World Bank.

Rolf, Arnold, and Joachim Munch. 1996. *Questions and Answers on the Dual System of Vocational Training in Germany*. Bonn: Federal Ministry of Education, Science, Research and Technology, Public Research Division, D-53170.

Appendix 6A Key Statistics on Students in Germany's Educational and Training System: 1996

Germany's Population	82,012,000
Male	39,955,000
Female	42,057,000
Persons in Employment	34,408,000
Male	20,043,000
Female	14,365,000
Unemployed Persons	3,965,000
Total Number of Students	12,551,000
At general schools	10,071,000
At vocational schools	2,480,000
Students at Higher Education (1994 Data)	1,857,000
At universities and college of education	1,412,000
At Fachhochschulen	445,000
Total Number of Apprentices	1,592,000
New training contracts	574,000
Apprentices According to Training Sector	
Industry and commerce	707,000
Crafts	627,000
Agriculture	34,000
Public sector	50,000
Independent professions	160,000
Home economics	13,000
Maritime shipping	3,000
Participants in Continuing Education Courses	5,223,000
On-the-job training	2,333,000
Courses offered by chambers	197,000
At special training centers	1,501,000
At a vocational school or higher education institution	732,000
Distance learning	71,000
Other	389,000

Source: Federal Ministry of Education, Science, Research and Technology, 1997a

Appendix

QUESTIONS FROM NATIONAL SCIENCE FOUNDATION INVITATIONAL MEETING, 1996: SCIENCE, MATHEMATICS, ENGINEERING, AND TECHNOLOGY EDUCATION PROGRAMS IN OTHER COUNTRIES

More than 30 representatives from higher education, professional associations, business and industry, government agencies, international organizations, and the public sector met in this two-and-a-half-day meeting to explore aspects of technician education and training that might influence U.S. productivity. One conclusion was that other countries with successful education and training systems have much to offer the United States. The following questions emerged for further study. They were shared with the writers who contributed to this monograph as a resource to help guide their chapter presentations.

The Technician Profession

Is there a job classification for technicians and are there formal or legal qualifications required to perform these jobs?

What functions are performed by these individuals?

Is there a close correlation between what Americans describe as technicians and the description used in the country you are characterizing?

What are the policies, procedures, and standards for certification of technicians?

Are technicians expected to be certified by some government or certifying agency? If so, who establishes and conducts the technician certification?

What are the upward mobility career opportunities for technicians?

Technician Education

What formal education or training programs are available for technician preparation?

What are the prerequisites, entry qualifications, nature of content, methods employed, exit examinations, trends, and issues of technician education?

What are some of the salient features of technician education programs; that is, what is the balance between technology overview and technically specialized content, the balance between lecture and laboratory, the nature of laboratory and field experience, etc.?

What mix of general education and technical courses is provided in the curriculum?

What is the equivalent academic level (high school; two-, three-, or four-year college, etc.) of technician education programs?

Is industrial training, apprenticeship, or cooperative education an integral component in these programs? If so, how are activities such as student placement, performance monitoring, and financial reward coordinated?

Are technician programs accredited by some organization? If so, what are the policies, procedures, and standards for accreditation?

What key resources (journals, books, videos, computer forums, etc.) are available to inform the discussion about technician education?

Who are the important and best-informed people in technician education in business and industry, labor, government, and education?

Partnerships

What are the partnerships (academic, government, labor, business and industry) to ensure effective technician education programs?

Supply and Demand of Technicians

What methods are used to ensure a sufficient supply of technicians to meet workforce needs, currently and in the future?

What methods are used to define and ensure appropriate attributes of technicians to meet industry needs?

What mechanisms or initiatives are undertaken to correct the ever-present fluctuations in the job market; that is, either an overabundance or a shortage of qualified technicians in the country?

Professional Development of Technicians

What formal education or training programs are available to technicians for continuing professional development?

What professional societies exist for technicians?

The Future

What are the issues, tensions, emerging trends, and new directions in technician education?

What are common concerns that warrant future cooperation, collaboration, and consultation among the principals concerned about the economy? Do these concerns trigger possible collaboration across national boundaries—for example, reciprocity in certification?

How are opportunities for collaborative programs and projects encouraged and supported?

Selected Bibliography

Abrahart, Alan, and Zafaris Tzannatos. 1996. *Australia: Confronting Institutional Impediments*. Washington D.C.: World Bank.

Alexim, João Carlos, et al. 1993. *Vocational Training in Latin America*. Berlin: European Centre for the Development of Vocational Training.

Barlow, Kerry, Anne Junor, and Carmel Spark. 1995. *Gender-Inclusive Guidelines for Curriculum Designers and Writers*. Victoria, Australia: Australian Commission for Training Curriculum Products.

Castro, Claudio de Moura, and Torkel Alfthan. 1992. *Five Training Models*. Geneva: Training Policies Branch, International Labour Organization.

Club D. 1988. *Databases in Vocational Education and Training: The European Scene*. Berlin: European Centre for the Development of Vocational Training.

Colombo Plan Staff College for Technician Education. 1993. *Technical and Vocational Education: Japan*. Manila, Philippines: Colombo Plan Staff College.

Curtain, Richard. 1993. *Recent Developments in Work and Skill Formation in Japan*. Clayton, Victoria, Australia: National Key Centre in Industrial Relations, Monash University.

Dupont, George. 1990. *Education and Training in Europe: Comparative Analysis of Dynamic Aspects of Education and Vocational Training, Flows of People and Flows of Funds*. Berlin: European Centre for the Development of Vocational Training.

European Centre for the Development of Vocational Training. 1988. *Exploratory Study of the Role and Activities of "Centres of Excellence" in the Textile Industry in Four EEC Member States*. Berlin: European Centre for the Development of Vocational Training.

———. 1990. *The Role of Social Partners in Vocational Education and Training Including Continuing Education and Training*. Berlin: European Centre for the Development of Vocational Training.

Fressura, Nicola. 1995. *Vocational Education and Training in Italy*. Thessaloniki: European Centre for the Development of Vocational Training.

Frideres-Poos, Jose. 1994. *Vocational Education and Training in Luxembourg*. Berlin: European Centre for the Development of Vocational Training.

Further Education Unit. 1992. "Some Lessons from Germany." *Australian Training Review* February: 34–35. London: Further Education Unit.

———. 1992. *Vocational Education and Training in Europe. A Four-Country Study in Four Employment Sectors*. London: Further Education Unit. ERIC Document Reproduction Service No. ED 353 473.

Gallaher, Leo. 1994. *Vocational Education and Training in Ireland*. Berlin: European Centre for the Development of Vocational Training.

Geers, Frederik. 1995. *Vocational Education and Training in Belgium*. Thessaloniki: European Centre for the Development of Vocational Training.

Gilardi, Rudiger von. 1989. *In-Firm Trainers of Young People in the Framework of the Dual Vocational Training System of the Federal Republic of Germany*. Berlin: European Centre for the Development of Vocational Training.

Great Britain Scottish Education Department. 1991. *Six Years on: Teaching, Learning and Assessment in National Certificate Programmes in Scottish Further Education Colleges*. Edinburgh, Scotland: Her Majesty's Stationery Office.

Grooting, Peter. 1993. *Training in Transition: Comparative Analysis and Proposals for the Modernization of Vocational Education and Training in Poland*. Berlin: European Centre for the Development of Vocational Training.

Hall, William C. 1995. *National Profiles in Technical and Vocational Education in Asia and the Pacific: Australia*. Bangkok: United Nations Educational, Scientific and Cultural Organization Principal Regional Office for Asia and the Pacific.

Hatton, Michael J. 1995. *Exemplary Training Models in Industrial Technology*. Gloucester, Ontario: Association of Canadian Community Colleges. ERIC Document Reproduction Service No. ED 383 358.

Honig, Benson, and Francisco Ramirez. 1996. *Technicians, Technical Education and Global Economic Development: A Cross-National Examination*. Washington D.C.: American Education Research Association. ERIC Document Reproduction Service No. ED 392 992.

Institute for Employment and Adult Education Research. 1996. *Vocational Education and Training in the Republic of Austria*. Thessaloniki: European Centre for the Development of Vocational Training.

International Project on Technical and Vocational Education. 1995. *National Profiles in Technical and Vocational Education in Asia and the Pacific Rim: Fiji*. Bangkok: United Nations Educational, Scientific and Cultural Organization Principal Regional Office for Asia and the Pacific. ERIC Document Reproduction Service No. ED 399 370.

———. 1995. *New Technologies of Training for Technical and Vocational Education. International Expert Group Meeting Final Report*. Paris: United Nations Educational, Scientific and Cultural Organization. ERIC Document Reproduction Service No. ED 399 366.

Iwamoto, Muneharu. 1994. *Case Study on Technical and Vocational Education in Japan. Case Studies on Technical and Vocational Education in Asia and the*

Pacific. Victoria, Australia: Royal Melbourne Institute of Technology. ERIC Document Reproduction Service No. ED 391 027.

Kearns, Peter. 1993. *Review of Research and Development Structures and Practices for Vocational Education, Training and Employment in Five OECD Countries*. Adelaide, Australia: National Centre for Vocational Education Research.

Kemnitzer, Susan Coady. 1997. *An Overview of Engineering Education in Japan*, Special Scientific Report #97-08. Arlington, Va.: National Science Foundation.

Koch, Richard. 1989. *Vocational Education in France: Structural problems and present efforts towards reform*. Berlin: European Centre for the Development of Vocational Training.

Kroenner, Hans, ed. 1989. *Innovative Methods of Technical and Vocational Education: Report of the UNSEEN International Symposium*. Bonn, Germany: Federal Ministry of Education and Science.

Mahoney, James R., ed. 1996. *Improving Science, Mathematics, Engineering, and Technology Instruction: Strategies for the Community College*. Washington, D.C.: Community College Press, American Association of Community Colleges.

Munch, Joachim. 1995. *Vocational Education and Training in the Federal Republic of Germany*. Berlin: European Centre for the Development of Vocational Training.

National Vocational Education & Training Centre for Vocational Education Research. 1996. *Industry-Led Training System*. Leabrook, Australia: National Vocational Education and Training Centre for Vocational Education Research.

Neves, A. 1993. *Evaluation of Vocational Training in a Regional Context. A Synthesis Report*. Berlin: European Centre for the Development of Vocational Training. ERIC Document Reproduction Service No. ED 372 274.

Nielsen, Soren P. 1995. *Vocational Education and Training in Denmark*. Berlin: European Centre for the Development of Vocational Training.

Nothdurft, William E. 1989. *Schoolworks: Reinventing Public Schools to Create the Workforce of the Future: Innovations in Education and Job Training in Sweden, West Germany, France, Great Britain and Philadelphia*. Washington D.C.: Brookings Institute.

New South Wales Board of Vocational Education and Training. 1995. *Quality Initiatives and Best Practice in Public Providers*. Sydney: Board of Vocational and Education Training.

Nuebler, Irmgard. 1992. *Limits to Change in Training Systems: The Case of Germany*. Geneva: Training Policies Branch, International Labour Organization.

Orr, Thomas, et al. 1995. *Serving Science and Technology: Five Programs Around the Globe*. Technical Report 95-5-001. Aizuwakamatsu, Japan: Center for Language Research, Aizu University. ERIC Document Reproduction Service No. ED 389 173.

Pair, Claude. 1994. *The Changing Role of Vocational and Technical Education and Training. Pathways and Participation in Vocational and Technical Education and Training. Synthesis Report.* Paris: Organization for Economic Cooperation and Development. ERIC Document Reproduction Service No. ED 387 645.

Pandit Sunderlal Sharma Central Institute of Vocational Education. 1995. *Vocational Education: Organisational and management alternatives. Papers presented at International Workshop on Organisational and Management Alternatives for Vocational Education within the Educational System.* Bhopal, India: Pandit Sunderlal Sharma Central Institute of Vocational Education.

Partee, Glenda. 1994. *Designing Quality Programs: International Lessons on Youth Employment Preparation.* College Park, Md.: Center for Learning and Competitveness, School of Public Affairs, University of Maryland.

Prais, S. J. 1995. *Productivity, Education and Training: An International Perspective.* National Institute of Economic and Social Research. Cambridge: Cambridge University Press.

Rachwalsky, Klaus. 1991. *Vocational Training—Investment for the Future: The Dual System of Vocational Training in the Federal Republic of Germany.* Cologne, Germany: Carl Duisberg Gesellschaft.

Ramoff, Andre, et al. 1994. *Vocational Education and Training in Central and Eastern Europe.* Luxembourg: Office for Official Publication of the European Communities, and Lanham, Md.: UNIPUB (distributor).

Royal Melbourne Institute of Technology. 1994. *The Development of Technical and Vocational Education for the Islamic Republic of Iran—A Case Study in Quality Improvement.* Victoria, Australia: Royal Melbourne Institute of Technology. ERIC Document Reproduction Service No. ED 391 026.

Schlicht, Michael. 1994. "The Development of Vocation Training in Central and Eastern Europe. German Experience in Demand." *Education and Science 1.* ERIC Document Reproduction Service No. ED 371 150. Victoria, Australia: Royal Melbourne Institute of Technology.

Seeland, Suzanne, and Lotte Valbjorn. 1991. *Equal Opportunities and Vocational Training: 13 Years on the Results of CEDEFOP's Programme for Women 1977–1990.* Berlin: European Centre for the Development of Vocational Training.

Sellin, Burkart. 1991. *Euroqualifications for All New EC Approaches and Programmes for the Vocational Training of Young People.* Berlin: European Centre for the Development of Vocational Training.

Sorensen, John Houman. 1988. *The Role of the Social Partners in Youth and Adult Vocational Education and Training in Denmark.* Berlin: European Centre for the Development of Vocational Training.

Stavrou, Stavros. 1995. *Vocational Education and Training in Greece.* Thessaloniki: European Centre for the Development of Vocational Training.

Takashi, Uematsu. 1995. *National Profiles in Technical and Vocational Education in Asia and the Pacific: Japan*. Bangkok: United Nations Educational, Scientific and Cultural Organization Principal Regional Office for Asia and the Pacific.

Technical and Further Education National Centre for Research and Development. 1989. *International Conference on Recent Research and Development in Vocational Education Conference Papers*. Vol 1. Payneham, NSW: Technical and Further Education National Centre for Research and Development. ERIC Document Reproduction Service No. ED 399 804.

Twining, John. 1993. *Vocational Education and Training in the United Kingdom*. Berlin: European Centre for the Development of Vocational Training.

United Nations Educational, Scientific and Cultural Organization. 1990. *Trends and Development of Technical and Vocational Education*. Paris: United Nations Education, Scientific and Cultural Organization.

United Nations Educational, Scientific and Cultural Organization Principal Regional Office for Asia and the Pacific. 1995. "Establishing Partnerships in Technical and Vocational Education: Co-operation Between Education Institutions and Enterprises in Technical and Vocational Education." Chapter 4 in *International Cooperation—Contributions of International and German Experts*. Berlin: United Nations Educational, Scientific and Cultural Organization. ERIC Document Reproduction Service No. ED 399 362.

———. 1996. "Technical and Vocational Education: Toward Economic and Policy Development in Japan." In *United Nations Educational, Scientific and Cultural Organization Regional Conference: Policy Development and Implementation of Technical and Vocational Education for Economic Development in Asia & Pacific Rim Proceedings*. Bangkok: United Nations Educational, Scientific and Cultural Organization Principal Regional Office for Asia and the Pacific.

———. 1996. Technical Education for the Hi-Tech Era. In *United Nations Educational, Scientific and Cultural Organization Regional Conference: Policy Development and Implementation of Technical and Vocational Education for Economic Development in Asia & Pacific Rim Proceedings*. Bangkok: United Nations Educational, Scientific and Cultural Organization Principal Regional Office for Asia and the Pacific.

Women's Education and Training Coordination Unit. 1996. *Young Women and Vocational Education and Training, A Compendium of Good Practices*. Granville, New South Wales: Women's Education and Training Unit, Technical and Further Education (TAFE).

Index

A

administrative structure, 6, 21–22

apprenticeship, 14–15, 84–86. *See also* German system

apprenticeship experiences, uneven, 12

assessment, 7–8, 12, 23–25, 90–91, 115

Australia: adult and community education, 92; approaches for teacher preparation, 92–93; assessment approaches and student achievement, 90–91; continuing education, 8; critical relationships in technical education system, 77; Framing the Future, 93; government investment, 99; industry investment, 98; industry relationships, 90; industry training advisory bodies (ITAB), 76, 86, 88–90; information collection, 99–100; *Kangan Report*, 73–75; literacy, 97–98; middle-level skills, 91–92; mission statement and manpower development, 75–76; modularized training programs, 11; National Strategy, 93–94; National Training Framework Committee (NTFC), 79, 91; National Training Framework (NTF), 78–80; online technology, 97; opportunities for employee continuing education, 91; qualifications, 95; registered training organizations (RTO), 80, 82, 84, 92, 97, 99; role of public and private employer and employee groups, 87–90; school role, 96–97; skills recognition, 95–96; system effectiveness, 93–94; teacher education, 9; technical and further education (TAFE), 91–92, 94, 96, 100; training facilities, 99; under-represented groups, 98–99; unrecognized training, 96; User Choice, 89–90; Vocational Education and Training (VET), 73–78, 80, 82, 87–90, 92–96, 98, 100

Australian curriculum structure, 82; Aeroskills training package, 86–87; learning experience structure in training packages, 83–84; local autonomy in adjusting, 11; national training packages (NTP), 82–83, 85–87; new apprenticeships, 84–86

Australian National Training Authority (ANTA), 87–90, 100

Australian Qualifications Framework (AQF), 74

Australian Recognition Framework (ARF), 78, 80, 82

Australian skill standards and qualifications, 80, 81

B

basic skills emphasis, 11

C

colleges, 22, 27–28, 39–41, 109, 112–113, 135–136. *See also* universities

continuing and further education, 8, 19–21, 39, 41, 48, 91–92, 94, 96, 100, 116–117, 120, 136, 146–149

courses, industry-specific, 8

curriculum, 6–7, 39, 41; basic electronics, 28–30; local autonomy in adjusting, 11; practical training, 7. *See also under specific countries*

curriculum structure, 82, 107–109

D

Danish Center for Technology-Supported Learning, 20

Danish Confederation of Trade Unions, 21

Danish educational system, 16; continuing education and training (AMU system), 19–21; credit transfer among institutions, 18; folkeskoles, 16–17; funding for, 26; goal of vocational education, 15; higher technical (HTX) *vs.* higher commercial (HHX) tracks, 17–18; national administrative structure, 21–22; secondary education, 17; system effectiveness and key issues, 29–32; technical training, 17–18; vocational education and training (VET), 17–18

Danish Employers' Confederation, 21

Danish Ministry of Education, 8, 20, 21

Danish Ministry of Labour, 8, 20, 21

Danish Technological Institute, 13

Denmark, 13; Act on Vocational Education and Training, 15; Adult Vocational Training System, 19–22; Apprenticeship Acts, 14–15; assessment, 23–25; basic electronics curriculum, 28–30; Basic Vocational Education Act, 15; continuing education, 8; Department of Vocational Education and Training, 21; Erhvervs Uddannelses-Syd (EUC-Syd) Technical College, 22, 27–28; mission of technical training program, 14–15; national-level trade committees, 22; program for electronics technicians, 27–30; relationship among partners in developing programs, 23, 24; roles of public and private sectors, 23; "sandwich" model for basic technical programs, 25–27; Slagteriskolen, 22; teacher education, 9; teacher preparation, 25; Technical College of Jutland, 22; technology-based learning, 20; unions, 14; Vocational Training Council, 21–22

disabled persons, training of, 12

E

education: importance of, 2; market competition and, 2; proportion of general and technical, 10–11; secondary, 17, 130–131, 157–158. *See also* technician development systems; *specific countries; specific types and aspects of education*

education ministries, 6, 8, 20, 21, 77, 99, 103, 131, 140

Electronic Component Technology Design, 58–59

electronics curriculum, 28–30

electronics industries, 36, 126

electronics technicians, 71; program for, 27–30

electronics technology programs, 138

engineering education, 105–107, 135–139

entrepreneurs *vs.* producers, 10

G

German system, 143, 156–157; access to, 145–146; adult and continuing education, 146–147; Bundesinstituts fur Berufsbildung (BIBB), 154–156; community adult education centers, 148; community and other public groups, 149; corporations, 148; *Fachhochschulen*, 143–145; higher education, 148; immediate organizations, 148–149; industrial commitment, 158; private trainers, 148; routes to higher education, 145; secondary school system and universities, 157–158; statistics on, 160; teacher education, 8–9; unions, 149; vocational training, 149–156; Vocational Training Act, 150–154; *Volkshochschulen* (VHS), 148

I

immigrants, training of, 12

industrial commitment, 158

industrial *vs.* sunrise economy, 36

industry: electronics, 36, 126; ongoing education in, 116–117; reduced involvement of, 9–10; technical training in, 112–113. *See also under specific countries*

industry investment, 98

industry needs, meeting, 47

industry relationships, 90

industry-specific courses, 8

industry training advisory body (ITAB), 76

Investment in People program, 8

Israel, 125, 140–141; continuing education, 136; Council for Higher Education (CHE), 132–133; economy, 125–129, 140; education system, 129–130; electronics industry, 126; electronics technology programs, 138; higher education, 132; investment in research and development, 127, 128; National Institute of Technological Training (NITT), 134–135, 137; Open University, 132–133, 137; Organization for Rehabilitation by Training (ORT), 125, 136; Pedagogy Center for Research and Development, 9; practical engineering program, 137–139; secondary schools, 130–131; technical education, 133–140; Technion, 129; Tel-Aviv University College of Practical Engineering (TAU-PE), 135–136; vocational and technical education, 131

J

Japan: controlled and inflexible system, 11; teacher education, 8

Japanese vocational and technical education system, 103; articulation to higher education and employment, 110–111; attracting students to technical careers, 121; colleges of technology, 109; Denso, 112, 115; encouragement of creativity, 120; entry requirement for associate degree programs, 114–115; exit requirements for students, 115; gender imbalances, 120–121; generalized and specialized courses, 108–109; Hitachi Ibaraki Technical College, 112–113; implications for American system, 122; issues for future, 121; lifelong learning, 116, 120; Monbusho, 103–106, 111, 113–116; Nippon Electric Technical Training College, 113; ongoing education in business and industry, 116–117; preparation and enhancement of teachers and faculty, 118–119; program effectiveness assessment, 115; public and private sector employer and employee groups, 111; school system, 104–105; specialization of institutions and tracking of students, 105–106; student recruitment, retention, and placement, 117–118; system effectiveness, 119–120; technical training in business and industry, 112–113; types of engineering and science technician education, 105–107; types of schools, 106–107

L

learning, assessing and crediting, 12. *See also* assessment

M

management information systems, need for better, 12

market competition, education and, 2

ministries of education, 6, 8, 20, 21, 77, 99, 103, 131, 140

N

National Labor Market Authority. *See* Denmark, Adult Vocational Training System

National Science Foundation (NSF) meeting, 1, 161

national systems, 4; strengths, 4–5; weaknesses, 5

O

online technology, 97

P

partnerships, 162

producers *vs.* entrepreneurs, 10

public information systems, 12

S

Scotland, 35; Association of Scottish Colleges, 40; continuing education, 8, 39, 41, 48; curriculum, 39, 41; Electronic Component Technology Design, 58–59; further education colleges (FE), 39, 41; Glasgow, 35–36; higher national units, 45; historical overview, 36–38; introduction of electronics industries, 36; Introduction to Semiconductor Applications, 66; Investors in People (IP), 47–48; lecturer training, 48; local autonomy in adjusting curricula, 11; local enterprise companies, 41; Local Enterprise Companies (LEC), 40; mission of technical education program, 38–39; modularized training programs, 11; National Certificate Module Descriptor, 66; National Certificate Programme, 42, 43; national certificates, 45; Progress of Electrical Technician, 71; Progression, 49; relationship among education partners, 39, 40; shift from industrial to sunrise economy, 36; Teacher Certificate in Further Education, 48; teacher education, 9; University for Industry (UFI), 49–50; Workstart Programme, 43–44, 64–65

Scottish Council for Education Technology (SCET), 46, 48, 50

Scottish Further Education Unit, 48

Scottish Office Education and Industry Department (SOEID), 40

Scottish Qualification Authority (SQA), 38–40, 42

Scottish Qualification Authority (SQA) Higher National Unit Specification, 52; Statement of Standards, 53–57

Scottish School of Further Education, 48

Scottish system and framework: flexibility, 46; meeting industry needs, 47; multicourse use, 46; multipurpose, 46; quality assurance, 47; transparency, 46, 47

Scottish Vocational Qualification (SVQ), 42, 49; General (gSVQ), 42–44, 49, 60–63

secondary education, 17, 130–131, 157–158

skill standards and qualifications, 80, 81

skills, middle-level, 91–92

skills recognition, 95–96

student achievement, 90–91

students: attracting to technical careers, 121; exit requirements for, 115; limited pathways for, 11–12; reduced involvement of, 9; tracking of, 105–106

T

teacher education and preparation, 8–9, 25, 48, 92–93, 118–119

technical competency, growing importance of, 2

technician development programs, 1; knowledge gained from, 1

technician development systems, 3; adjusting to rapid change, 11; administrative structure, 6; guiding principles, 3; national approaches, 4–5. problems and concerns addressed, 5–6; similarities of concerns among countries, 5–6; similarities of systems, 3; trends and problems, 9–12. *See also* national systems; *specific topics and specific countries*

technician profession, 161

technicians: professional development, 163; supply and demand of, 163. *See also specific topics*

trade, international, 2

training. *See* education; vocational education and training

U

unions, 10, 12, 14, 21, 149; conflict between employers and, 12; declining power of, 10

United States: implications of Japanese system for, 122

universities, 49–50, 132–133, 135–137, 157–158. *See also* colleges

V

vocational education and training (VET), 17–18, 131, 149–156; Australian, 73–78, 80, 82, 87–90, 92–96, 98, 100. *See also* Japanese vocational and technical education system

W

worker mobility, as advantage and liability, 10

About the Editors

James R. Mahoney is the former director of academic, student, and international services at the American Association of Community Colleges. He now serves as a consultant to the association on its work with the National Science Foundation.

Lynn Barnett is the director of academic, student, and community development at the American Association of Community Colleges. She is the principal investigator for a science, math, engineering, and technology project funded by the National Science Foundation.